THE WEATHER OF THE PACIFIC NORTHWEST

Cliff Mass

THE WEATHER
OF THE PACIFIC NORTHWEST

A SAMUEL AND ALTHEA STROUM BOOK

UNIVERSITY OF WASHINGTON PRESS *Seattle & London*

This book is published with the assistance
of a grant from the Samuel and Althea Stroum
Endowed Book Fund.

UNIVERSITY OF WASHINGTON PRESS
PO Box 50096, Seattle, WA 98145, USA
www.washington.edu/uwpress

LIBRARY OF CONGRESS
CATALOGING-IN-PUBLICATION DATA
Mass, Cliff.
The weather of the Pacific Northwest / Cliff Mass.
p.   cm.
Includes bibliographical references and index.
ISBN 978-0-295-98847-4 (pbk. : alk. paper)
1. Northwest, Pacific—Climate.   2. Weather
forecasting—Northwest, Pacific.   I. Title.
QC984.N97M37 2009
551.69795—dc22   2008025873

CONTENTS

ACKNOWLEDGMENTS

Many individuals have greatly aided the writing and preparation of this book. Beth Tully drafted most of the drawings and schematics, using her exceptional skills for communicating difficult concepts and complex phenomena. Professor Richard Reed, University of Washington Research Meteorologist Mark Albright, and National Weather Service Area Manager Brad Colman taught me much about the intricacies of Northwest weather. I have also learned a great deal about Northwest meteorology working with graduate students over the years, including Justin Sharp, Jim Steenburgh, Garth Ferber, Fang Chien, Matt Garvert, Brian Colle, Eric Grimit, Tony Eckel, and Bri Dotson. Several past and present research meteorologists in the department—David Ovens, Richard Steed, Jeff Baars, and Ken Westrick—have been my con-stant colleagues in dissecting the inner workings of regional weather features. The Seattle National Weather Service office has provided local weather information and data, as well as an active and productive interaction that have greatly enhanced several chapters. This manuscript has been sub-stantially improved based on the comments of Brad Colman, Mark Albright, David Laskin, Beth Fuget, Justin Sharp, Steve Todd, and many reviewers in my department. Art Rangno provided many of the excellent cloud photographs, and University of Washington Press copyeditor Julie Van Pelt greatly improved the book's readability. The attractive design of the book was created by Ashley Saleeba, and UW Press editors Michael Duckworth and Mary Ribesky were highly sup-portive during the several years it took to bring this project to fruition. The continued interest of listeners to my weekly radio segment on KUOW's *Weekday* program has been an important inspiration for this book, and much of the material herein answers frequent audience questions. Finally, I acknowledge the financial support of the Northwest Modeling Consortium, the National Science Foundation, the National Weather Service, and the U.S. Navy's Office of Naval Research, which made possible much of the modeling and research that provides the intellectual foundation for this book.

THE WEATHER OF THE PACIFIC NORTHWEST

1

THE EXTRAORDINARY WEATHER
OF THE PACIFIC NORTHWEST

||

ON COLUMBUS DAY 1962, THE STRONGEST NONTROPICAL CYCLONE TO STRIKE

the continental United States during the past one hundred years pummeled the West Coast

from northern California to southern British Columbia. With winds gusting to nearly 200 miles

per hour over the coastal headlands and 100 miles per hour around Puget Sound and the

northern Willamette Valley, the damage was devastating (figure 1.1). Power outages extended

over most of the region, more than fifty thousand buildings were damaged, and forty-six

people lost their lives. How could such a storm strike a region known for its benign weather

and the velvet softness of its clouds, fog, and incessant light rain? Northwest weather is often

surprising, both in its intensity and in the startling contrasts between nearby locations.

1.1. On the state capitol grounds in Salem, Oregon, the bronze statue *The Circuit Rider* was toppled by hurricane-force winds during the 1962 Columbus Day Storm. Photo by Hugh Stryker and provided courtesy of the Salem Public Library Historic Photograph Collections.

The weather of the Pacific Northwest is exceptional in many ways. While much of the eastern two-thirds of the United States endures warm, humid summers and cold, often snowy winters, the western side of the Northwest enjoys mild, dry summers and temperate, wet winters. In much of the country, weather varies gradually from one location to another; in contrast, rapid changes and localized weather are the norm in the Northwest, where radically different weather conditions are often separated by a few miles. While thunderstorms are a major feature of the weather in most of the country, in the Northwest they are infrequent events, with strong thunderstorms, tornadoes, or hail a rarity. Although hurricanes can strike the Atlantic and Gulf coasts of the United States and greatly influence the weather far inland, the Northwest is never affected by such tropical storms. Not only is the weather different in the Northwest, but so is its prediction. While weather forecasts for the central and eastern portions of the nation are enhanced by the dense observing network over North America, Pacific Northwest predictions are degraded by the sparsity of observations to the west, since nearly all West Coast storms originate over the relatively data-poor North Pacific.

Although Northwest weather is usually gentle and benign, some of the most severe weather of the continent is experienced here. Intense Pacific low-pressure systems, like the Columbus Day Storm, packing hurricane-force winds and extending over considerably larger areas than tropical storms, can bring destruction to wide swaths of the region. Localized windstorms, often associated with gaps in the high Northwest mountains or air flowing across major terrain features, have produced severe small-scale winds reaching 100 miles per hour or more. One such event destroyed the Hood Canal Bridge in 1979 at a cost of over 140 million dollars, and others have peeled off roofs in the Cascade foothills town of Enumclaw. While snow is infrequent and generally light over the Northwest lowlands, the heaviest measured snowfall in the . world strikes the Cascade Mountains, resulting in buried roads and avalanches. During the last week of December 1996, such heavy snow closed all Cascade passes in Washington and resulted in widespread building collapses on both sides of the mountains. Although rainfall amounts are usually light to moderate during Northwest winters, Pineapple Express rainstorms, associated with rivers of atmospheric moisture originating north of Hawaii, can bring several feet of rain to Northwest communities over a few days, resulting in catastrophic flooding and mudslides. Such conditions hit the region with full force during November 2006, with Mount Rainier National Park experiencing the most severe damage since its inception (figure 1.2) and losses from flooding in Oregon and

Washington totaling hundreds of millions of dollars. Billion-dollar storms have occurred several times in the Northwest since 1980 and all of them have been associated with severe flooding.

Startling weather contrasts over small distances are some of the most singular aspects of Northwest weather. The high terrain of the region often separates radically different climate and weather regimes, with transitions occurring over a matter of miles. The Olympic Mountains are a prime example: rain-forest conditions and annual precipitation approaching 200 inches a year are found on its western slopes, such as within the Hoh River valley, while a few dozen miles away, on the mountains' northeastern side, Sequim typically receives about 15 inches a year. It is easy to see why the latter is a magnet for retirees in search of California-like conditions in the Northwest. Large contrasts similarly occur over the Columbia Gorge, with the change from the wet, lush forest environment near Cascade Locks to arid, barren conditions just east of Hood River occurring in a little over 20 miles and less than a half hour's drive on Interstate 84. On December 18, 1990, an unexpected foot

1.2. Record-breaking rains on November 6, 2006, caused catastrophic slope and road failures across Mount Rainier National Park, resulting in the closure of much of the park for months. This picture shows damage to Nisqually Road at Sunshine Point in the southwestern portion of the park. Photo courtesy of the National Park Service.

of snow crippled the city of Seattle during rush hour, while 20 miles to the north and south the ground remained bare. Northwest winds can also vary greatly over short distances. An extreme case occurred on the night of December 24, 1983, during a severe cold spell over the region. Air rushed westward through a gap in the Cascades and descended toward Enumclaw and vicinity, bringing wind gusts of over 120 miles per hour that tore off roofs and crumpled high-tension power-line towers. In contrast, 25 miles to the northwest in Seattle the winds were dead calm. No wonder local TV stations love to describe Northwest weather as "weird" or "wacky."

The Northwest is also home to notable weather anomalies. When air descends the steep terrain that bestrides the Oregon-California border (the Siskiyou/Klamath Mountains), the southern Oregon coast can be 10–20 °F warmer than the rest of the Northwest, with high temperatures soaring into the 80s °F even in midwinter. Not surprisingly, the local chamber of commerce advertises this area as the "banana belt" of the Northwest. Other Northwest locations are famous for their extreme cold. Mazama and Winthrop, Washington, located in a deep valley protruding into the northern Cascades, are often the coldest locations in the state, both setting the all-time record low for Washington of –48 °F on December 30, 1968. Even colder temperatures can occur within the frigid valleys of the uplands of eastern Oregon, where Ukiah and Seneca cooled to –54 °F during the winter of 1933. The all-time record for annual snowfall in the world is held by the Mount Baker Ski Area, where 1,140 inches fell during the 1998–99 winter season, breaking the previous world record (1,122 inches) at Mount Rainier. The snow was so plentiful that

year that skiing had to be suspended until the ski lifts were dug out.

The Northwest is also home to what might be called weather curiosities. Air streaming over the Northwest mountains can sometimes create wave-like clouds that resemble hovering flying saucers; in fact, such an apparition set off the UFO craze in 1947. Changes in air temperature above Puget Sound often cause optical effects in which ferry boats and other marine vessels appear to be flying above the water and shorelines seem thrust high into the skies. More ominously, the combination of strong winds and arid conditions east of the Cascades can produce terrible dust storms that decrease visibility to near zero and cause multicar accidents. The eruption of Mount Saint Helens in 1980 covered vast areas of the Northwest with darkness, with the ash cloud acting as an insulator that kept temperatures virtually constant for over twelve hours across much of eastern Washington. And the foggiest location in the continental United States is found near the outlet of the Columbia River at Cape Disappointment, where the typical year brings 106 days of dense fog with a visibility of a quarter mile or less.

Serious misconceptions about Northwest weather abound and many are put to rest in these pages. Probably the most repeated unsubstantiated claim is that Seattle receives more rain than virtually anywhere else in the continental United States. Not true. With an average annual precipitation of roughly 37 inches, Seattle's rainfall is handily beaten by New York City (47 inches), Miami (56 inches), and many other locations across the eastern, central, and southern portions of the country. Another canard is that the Northwest is wet year-round. The truth is that Northwest precipitation

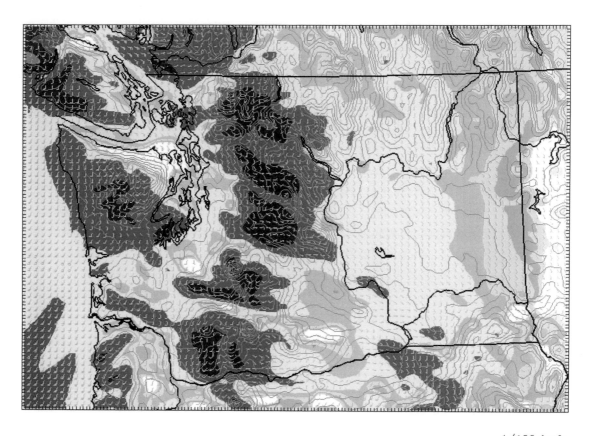

1.3. High-resolution computer predictions of Northwest weather are now greatly improving forecasts. This graphic shows a thirty-six-hour forecast of precipitation over Washington State using a state-of-the-art computer-forecasting model. The values shown are for the three hours ending at 4:00 AM on January 15, 2006, and are in hundredths of an inch, with blue and dark green indicating the heaviest precipitation. Note the rain shadow to the northeast of the Olympic Peninsula and the heavy rainfall over the southwestern side of the Olympic Mountains and the western slopes of the Cascades. Terrain contours and wind flags are also shown.

is concentrated in relatively few months from November through February and that our summers are among the driest in the nation—even including the desert Southwest. Finally, some assert that Northwest mountains make weather prediction difficult; as explained later, the mountains have the opposite effect, improving forecast skill and giving Northwest forecasters advantages over their eastern colleagues.

Northwest meteorologists are often the brunt of local humor, and it is not unusual to hear people muse that dice would be a more reliable forecast guide. But the truth is that forecasts *are* getting better. Making use of new technologies—such as weather radar, satellite imagery, and high-resolution computer weather simulations—meteorologists have unraveled many of the details of Northwest weather, and forecasting skill has increased substantially (figure 1.3). While in decades past, major windstorms like the Columbus Day Storm of 1962 were poorly predicted, many of the recent great blows, such as the Inauguration Day Storm of 1993 or the Hanukkah Eve Storm of 2006, were forecast accurately days in advance. Something has changed, and this book describes the evolving technologies that have made improved predictions possible.

With its dependence on melting snow as a source of water and hydroelectric power during the summer and early fall, the Pacific Northwest may be particularly sensitive to the effects of global warming. Although the mountains and complex land-water contrasts of the Northwest make prediction of its future climate challenging, recent scientific advances are slowly revealing the region's future. Some of these revelations are surprising, including an increase in springtime clouds west of the Cascades and local warming "hot spots." As described in this book, the effects of global warming will vary greatly across the region, with warming weakened near the coast and enhanced on mountain slopes.

Both poorly understood and forecast until recently, the complex meteorology of the Northwest has been the subject of intense scrutiny by local weather scientists since the late 1970s. Making use of these insights, this book describes the weather of the region stretching from southern British Columbia to the California border, and from the western slopes of the Rockies to the Pacific Ocean. The goal is to provide a description of Northwest weather that is both accessible to a layperson and scientifically accurate.

2

THE BASICS OF
PACIFIC NORTHWEST WEATHER

IF ONE COULD USE A SINGLE PHRASE TO DESCRIBE PACIFIC NORTHWEST

weather, "wet and mild" would be a start, but not a particularly exact one. Although the

region west of the Cascade crest is considered "wet" by many, it enjoys some of the driest

summers in the nation and receives less annual precipitation than much of the eastern United

States. East of the Cascades, where arid conditions dominate, "wet" is certainly not an apt

description, and east-side temperature extremes, ranging from −48 to 119 °F, makes "mild"

a misnomer at times. Northwest weather and climate are dominated by two main elements:

(1) the vast Pacific Ocean to the west and (2) the region's mountain ranges that block and

deflect low-level air. Together, these factors explain many of the dominant and fascinating aspects of the region's weather. The ocean moderates the air temperatures year-round and serves as a source of moisture, and the mountains modify precipitation patterns and prevent the entrance of wintertime cold air from the continental interior.

WHY IS PACIFIC NORTHWEST WEATHER GENERALLY MILD?

The Pacific Northwest is located in the northern hemisphere midlatitudes, a zone stretching from approximately 30° to 60° north where winds generally blow from west to east. This eastward movement of air is usually not uniform in strength, but is typically strongest in a relatively long, narrow current a few hundred miles across and a few miles deep, known as the *jet stream*. Usually centered 5–8 miles above the surface, jet-stream winds often reach 100 to 200 miles per hour during the winter. Weather systems, such as the low-pressure systems that bring rain and wind, tend to follow the jet stream, and thus the jet stream can be considered an atmospheric "highway" for storms and precipitation. The jet stream undulates north and south like a sinuous snake but is not continuous around the globe, since there are longitudes where it is broken or weak.

Before reaching the Pacific Northwest, the eastward-moving air traverses thousands of miles of the Pacific Ocean. Crossing the ocean over a period of several days, the air near the surface is profoundly modified, moistening and taking on the temperature of the underlying ocean surface. The surface temperature of the midlatitude northern Pacific Ocean is relatively temperate even during the winter, typically ranging from 45 to 50 °F between Japan and the Northwest coast (figure 2.1). Thus, low-level air reaching the Pacific Northwest

during the winter is generally mild and moist, resulting in typical wintertime air temperatures west of the Cascades rising into the mid-40s. The vast Pacific Ocean, like a huge liquid flywheel, only warms slowly during the summer. Thus, sea-surface temperatures off the Northwest coast vary little during the year and rarely rise above the lower 50s °F.

As the jet stream and associated storms weaken and retreat northward during the warm season, high pressure builds northward over the eastern Pacific (see figure 2.11 later in this chapter). With higher pressure offshore, cool air from the ocean is pushed inland, ensuring that summertime temperatures west of the Cascades remain moderate, rarely exceeding 90 °F along the coast and over the Puget Sound lowlands. Only when the wind direction reverses and air moves westward from the warm continental interior can temperatures reach the upper 80s °F and beyond over the western side of the Cascades.

The other major element of Northwest weather is the terrain, ranging from the formidable Rocky and Cascade mountains, which reach 5,000 to 14,000 feet, to the low coastal mountains, which attain only 3,000 or 4,000 feet (figure 2.2). East of the Cascades, a topographical "bowl" encompasses the lower Columbia valley, including the Tri-Cities in Washington and Pendleton, Oregon, while eastern Oregon is an elevated plateau, with some higher peaks and several major valleys.

In the winter, the Rockies and Cascades form a double barrier to the cold air of the continental interior (figure 2.3). The Rockies act as the Northwest's first line of defense, blocking the cold air that develops over the snowfields of the Canadian Arctic and that subsequently moves southward into the interior of the continent. If the cold air becomes deep enough, some can push over the Rockies, but since air warms as it descends, the air moving down the western slopes of the Rockies reaches eastern Washington and Oregon considerably warmer than air at similar elevations east of the continental divide. Next come the Cascades, which block the westward movement

of most of the cold, dense air that does manage to reach eastern Washington and Oregon. During the unusual circumstances when the cold air east of the Cascade crest becomes deep enough to push westward across these mountains, it is warmed further as it descends the western side. In short, because of the blocking effects of the Rockies and Cascades, eastern Montana is colder than eastern

SEA-SURFACE TEMPERATURE

2.1. Climatological sea-surface temperatures (°F) during late December and late July. The sea-surface temperatures of the northeastern Pacific west of the Northwest remain in the mid-40s to the mid-50s °F year-round. Image courtesy of the U.S. Navy's Fleet Numerical Meteorology and Oceanography Center, Monterey, California.

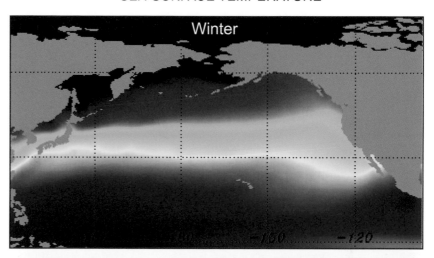

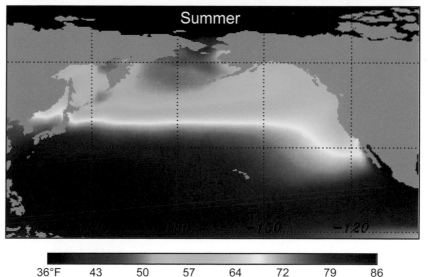

36°F   43   50   57   64   72   79   86

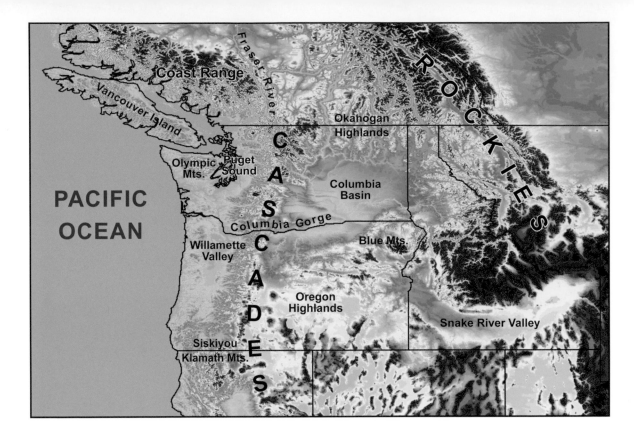

2.2. Color-enhanced topographic map of the Pacific Northwest.

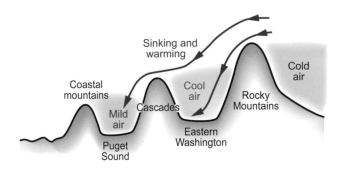

2.3. The major mountain ranges of the Northwest protect the region from the frigid air of the interior of North America. At low levels, the coldest air is found east of the Rockies, within the continental interior. Air that makes it across the Rockies warms as it descends into eastern Washington and Oregon. Any air that crosses the Cascades is further warmed as it descends over the western slopes and is compressed by the higher pressure at lower elevations. Illustration by Beth Tully/Tully Graphics.

Washington and Oregon, which in turn are colder than western Washington and Oregon. Even the most successful football coach would be impressed by the Northwest's multilayer defense against the cold-air opposition.

Although the Rockies and Cascades usually prevent frigid, Arctic air from entering western Oregon and Washington, two major gaps or weaknesses in the Cascades permit the entrance of cold air under the proper circumstances. The first, commonly called the Fraser Gap, follows the Fraser River valley from the interior of British Columbia to its terminus northeast of Bellingham (see figure 2.2). The second is the narrow Columbia River gorge, which provides a near sea-level westward passage for cold air originating in eastern Washington.

Cold air moves southwestward down the Fraser Gap when cold, Arctic air over the Yukon and northern British Columbia deepens sufficiently to push into the interior of British Columbia.

Subsequently, the cold flow follows the Fraser River valley westward since the valley is the lowest conduit across the Canadian Coast Mountains (the northward extension of the Cascades into British Columbia). Similarly, cold air often moves westward though the Columbia Gorge when higher pressure and cold air become entrenched over eastern Washington. Strong winds can develop over the western portions of both the Fraser Valley and the Columbia Gorge during such cold-air outbreaks as air accelerates between the high pressure east of the mountains and lower pressure to the west. As described in chapter 4, cold air from the Fraser Gap is often associated with western Washington snowstorms, while cold air in the Columbia Gorge can produce snow and ice storms over the Portland metropolitan region.

A SURVEY OF PACIFIC NORTHWEST TEMPERATURES

Temperatures across the Pacific Northwest are controlled by proximity to water and by elevation, the amount of clouds, and the position of major mountain barriers. Figure 2.4 illustrates the typical surface air temperatures[1] over the region for summer (July) and winter (January). During January, nighttime low temperatures west of the Cascade crest typically drop into the lower 30s °F, except for the coastal zone and areas near Puget Sound where cooling is tempered by proximity to relatively warm water. Somewhat cooler temperatures (upper 20s °F) extend eastward from the Columbia

Gorge toward Pendleton and the Tri-Cities. East of the Cascades, temperatures drop as elevation increases, with the coldest temperatures over the high terrain of northeast Washington and central Oregon, where nighttime temperatures typically plummet into the mid-teens. January maximum temperatures follow a similar, but warmer, pattern. Over the western Washington/Oregon lowlands, January high temperatures rise into the mid-40s and lower 50s °F during the day, with the warmest temperatures over the southern Oregon coast. In contrast, over the higher terrain of the Cascades, northeast Washington, and the central highlands of eastern Oregon, high temperatures generally remain well below freezing.

Clouds play an important role in producing the winter temperature distribution. West of the Cascades, incessant wintertime clouds reduce the maximum temperatures and increase the daily lows. During the day, clouds reflect a great deal of incoming solar radiation, which is why they look white in visible weather-satellite pictures shown on television. Reflecting the sun's rays back into space produces cooling. In contrast, clouds can warm the surface at night, since they lessen the ability of the ground to emit infrared radiation to space. Thus, cloudy nights generally are warmer than clear ones, and the low temperatures in cloudy western Washington rarely drop much below freezing. Interestingly, clouds also explain why winter low temperatures are often relatively high in the low-elevation bowl of eastern Washington, since the persistent winter low clouds of this area mitigate nighttime cooling.

Summer brings not only much warmer temperatures, but a very different pattern of temperature variation across the region. July minimum temperatures

---

1   Surface air temperatures are generally measured in shade at 2 meters, or roughly 6.5 feet above the ground.

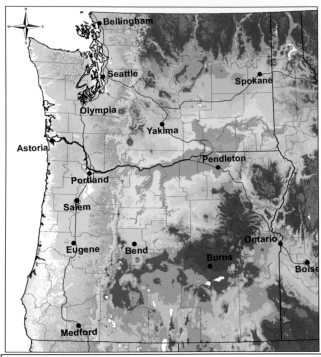

January Mean Minimum Temperature (deg F)

< -4 -4 - -2 -2 - 5 5 - 8 8 - 12 12 - 14 14 - 16 16 - 18 18 - 20 20 - 22 22 - 24 24 - 26 26 - 28 28 - 29 29 - 31 31 - 32 32 - 34 34 - 36 36 - 38 38 - 40 40 - 44 44 - 48 48 - 51 51 - 55 > 55

0 10 20 40 60 80 100 120 140 160
Miles

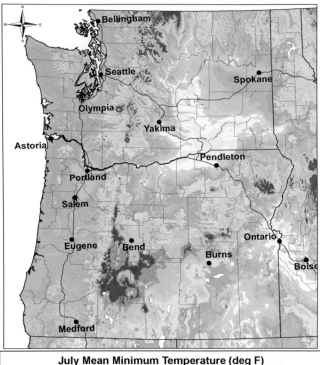

July Mean Minimum Temperature (deg F)

< 20 20 - 23 23 - 26 26 - 30 30 - 33 33 - 35 35 - 38 38 - 40 40 - 42 42 - 44 44 - 46 46 - 48 48 - 50 50 - 51 51 - 52 52 - 54 54 - 56 56 - 58 58 - 60 60 - 62 62 - 66 66 - 70 70 - 74 74 - 78 78 - 82 > 82

0 10 20 40 60 80 100 120 140 160
Miles

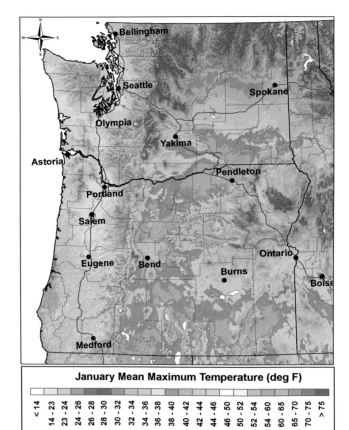

January Mean Maximum Temperature (deg F)

< 14 14 - 23 23 - 24 24 - 26 26 - 28 28 - 30 30 - 32 32 - 34 34 - 36 36 - 38 38 - 40 40 - 42 42 - 44 44 - 46 46 - 50 50 - 52 52 - 54 54 - 60 60 - 65 65 - 70 70 - 75 > 75

0 10 20 40 60 80 100 120 140 160
Miles

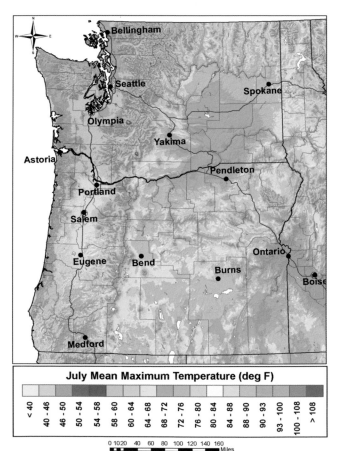

July Mean Maximum Temperature (deg F)

< 40 40 - 46 46 - 50 50 - 54 54 - 58 58 - 60 60 - 64 64 - 68 68 - 72 72 - 76 76 - 80 80 - 84 84 - 88 88 - 90 90 - 93 93 - 100 100 - 108 > 108

0 10 20 40 60 80 100 120 140 160
Miles

are relatively uniform west of the Cascades, with lows in the mid- to lower 50s °F. Warmer temperatures are found east of the Cascades in the lower elevations of the Columbia River basin, particularly in the topographic bowl encompassing the Tri-Cities and Pendleton, where temperatures only decline to about 60 °F. Over the higher elevations of the Cascades and the highlands of eastern Oregon, nighttime temperatures are chilly even in midsummer, with typical minimum temperatures in the 40s °F. For maximum summer temperatures, there are significant variations over the Willamette Valley, ranging from the 80s °F to the north to the 90s °F to the south; in contrast, over the western Washington lowlands, temperatures reach only into the mid-70s °F. These temperature differences are caused by terrain and proximity to water. While the western Washington lowlands are flooded with air from Puget Sound, the Straits of Juan de Fuca and Georgia, and the Pacific Ocean, the Willamette Valley is landlocked on three sides, limiting access to air tempered by a cool water surface. Thus, while air conditioning is rarely needed in most Puget Sound communities, it is often used in homes from Portland to Eugene. The Medford, Oregon, area, found in a topographic low spot in the middle of the Siskiyou/Klamath Mountains, is completely cut off from marine air and typically warms into the 90s °F during summer afternoons. The region's warmest temperatures are generally found over the lower elevations east of the Cascades near the

Columbia River, where temperatures frequently reach the mid-90s. As described in chapter 9, the all-time high temperature records for the Northwest have occurred in this heated bowl. Maximum temperature decreases with elevation over the Oregon plateau and the Cascade Mountains, with the lowest maximum temperatures in the upper 50s °F over the highest terrain.

Examining the annual variation in temperature around the Northwest reveals some intriguing differences (figure 2.5). Perhaps most striking is that the range of annual temperature is *far* larger east of the Cascades than over the more temperate western side. For example, average daily maximum temperatures typically vary by 60 °F between January and July east of the Cascades, while on the western side a 30-degree variation is usual. In Washington State, Seattle and the Quillayute weather station, on the northwest coast, experience nearly identical maximum temperatures (mid-40s) during the winter, with Seattle warming up about 10 °F more than Quillayute during midsummer. For both, the coldest temperatures occur during early January, followed by slow warm-up to the annual peak around August 1. East of the Cascades, Spokane is decidedly cooler than Yakima throughout the year (by about 5 °F), with the warmest temperatures sharply peaking near August 1 and lowest temperatures occurring around New Year's.

Oregon locations also tend to have their highest temperatures around August 1, except along the coast, where the warmest temperatures occur about a month later. North Bend is not the place to go for temperature extremes, with highs ranging from the low 50s °F in winter to mid-60s during August and September. The reason, of course, is the nearby Pacific Ocean, whose temperature only

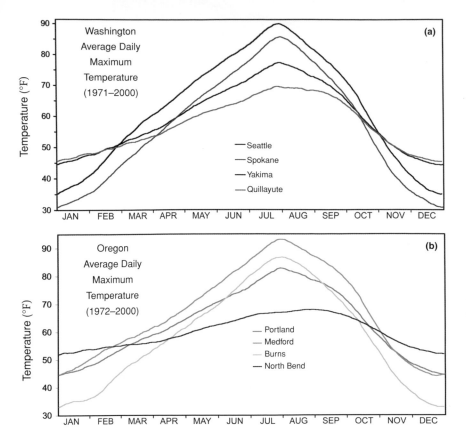

2.5. Average daily maximum temperatures for some locations in (a) Washington and (b) Oregon. Most stations in these states achieve their highest temperatures near August 1 and their lowest around January 1.

cools about 5 °F during the winter. In contrast, Burns, in the eastern highlands of the state, has a far greater temperature range, with highs near freezing in late December and January and the mid-80s °F during midsummer. Although similar to Portland during the fall and early winter, Medford gets far warmer in the summer, with average highs in the mid-90s—perfect for a wet run down the nearby Rogue River.

RAIN FOREST AND DESERT: WHY PRECIPITATION VARIES SO MUCH ACROSS THE PACIFIC NORTHWEST

Nowhere in North America are precipitation contrasts greater than in the Pacific Northwest. Driving east on Interstate 84 through the Columbia River gorge, one transitions from rain-forest conditions near Cascade Locks (80 inches per year) on the western side to an arid environment in The Dalles (13 inches), only 45 miles to the east (figure 2.6b). On the southwest side of the Olympics there is the sodden Hoh rain forest, which receives 140–160 inches a year, while 40 miles to the northeast the town of Sequim in the Olympic rain shadow enjoys a relatively dry, sunny climate with about 15 inches a year (figure 2.6a). As described in chapter 10, Sequim is so dry that some cacti grow there and irrigation is required for most crops.

The distribution of precipitation[2] over the Pacific Northwest is greatly influenced by the region's mountain ranges (figure 2.7). As explained

2   The term precipitation means the total of all types of water falling from clouds in both liquid or solid forms (rain, snow, ice, hail, and others).

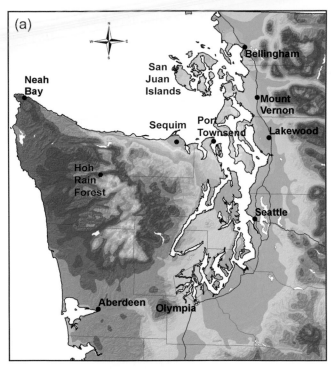

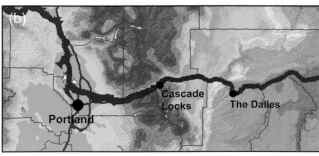

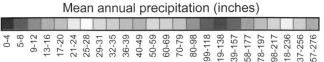

Mean annual precipitation (inches)

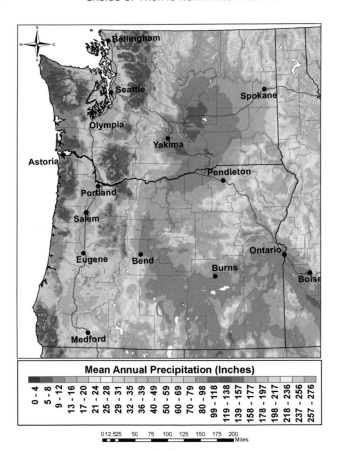

2.7. Annual precipitation (in inches) for Washington and Oregon. Heavier precipitation falls on the windward (typically western) slopes of the Cascades and coastal mountains, with arid conditions over the rain-shadow regions of eastern Washington and Oregon. Graphics courtesy of Chris Daly and Mike Halbleib of the Oregon State University PRISM group.

2.6. Annual total precipitation (1971–2000) over northwest Washington State (a) and the Columbia River gorge area of Oregon (b). Graphics courtesy of Chris Daly and Mike Halbleib of the Oregon State University PRISM group.

later in this chapter, clouds and precipitation are associated with rising air, while clearing occurs as air descends. Since air typically moves across the region from the southwest to the northeast during the wet, winter season, precipitation generally increases on the southwestern or western slopes of Northwest mountains where air is forced to rise. In contrast, precipitation typically decreases over the northeastern or eastern slopes where air descends. Meteorologists typically refer to the slope facing the wind as the *windward* side, while the drier, downwind slope where the air descends is known as the *leeward* side.

Moist air moving eastward from the Pacific Ocean is first forced to rise by the coastal mountains of Oregon, Washington, and Vancouver Island, producing substantial precipitation on their western slopes. Annual totals range from 60 to 80 inches on the lower coastal mountains to 90 to 160 inches on the western slopes of the highest coastal terrain, the Olympics. As the air crosses the coastal mountains and descends into the lowlands of western Washington and Oregon (Puget Sound through the Willamette Valley), the annual precipitation decreases to around 35 to 45 inches a year. Thus, the lowland urban corridor stretching from Bellingham to Eugene is in the rain shadow of the coastal mountains and enjoys a far drier climate than the coastal zone. If Lewis and Clark had known this, perhaps they would have built their encampment in the relatively dry Willamette Valley rather than the extraordinarily wet coastal location they chose (near Astoria, Oregon). In westside locations where the air frequently descends from higher terrain, such as northeast of the Olympic Mountains, precipitation is further reduced. As noted earlier, the town of Sequim lies in the middle of the Olympic rain shadow and receives only about 15 inches of precipitation a year, a total more typical of southern California. Another arid region is found in southwestern Oregon near Medford and Ashland, which are located in a bowl-like valley within the relatively high Siskiyou/Klamath Mountains that stretch from the coast to the southern Cascades. Medford typically receives only 18 inches of precipitation a year and is surrounded by rangeland, an environment more typical of the eastern side of the Cascades.

As air continues to move eastward it is forced to rise by the Cascades, and 60–120 inches of precipitation typically fall each year from southern Oregon into southern British Columbia on the barrier's western slopes. Although precipitation initially increases with elevation over the western side of the Cascades, it appears that precipitation begins to drop off at the highest elevations, above approximately 7,000 feet. Why? At such heights there is less blocking terrain to push the air upward, and the normal decrease of water vapor with height results in less precipitation formation. After crossing the Cascade crest, air descends rapidly over the eastern slopes of the Cascades, producing a sharp decrease of clouds and precipitation. Air descending into the bowl of eastern Washington produces extreme aridity, with annual precipitation decreasing to less than 10 inches a year. In contrast, eastern Oregon is mainly high plateau, so air subsides less on the eastern side of that state. Thus, eastern Oregon is dry, with annual totals of 10–20 inches, but less arid than eastern Washington. The weather-satellite image in figure 2.8 illustrates the enhancement of clouds west of the Cascade crest, with rapid evaporation to the east.

Infrequently, the region's winds blow from the east. On such occasions, the distributions of clouds and precipitation make a corresponding shift, with the eastern slopes of the Cascades and Olympics becoming enshrouded in clouds and showers, while the usually wet western slopes turn warm and dry. The mountain crests remain in cloud for either wind direction. That is why ski areas near the Cascade crest are often cloudy: they are in cloud whether the winds are from the east or the west. When the winds are from the southeast, the

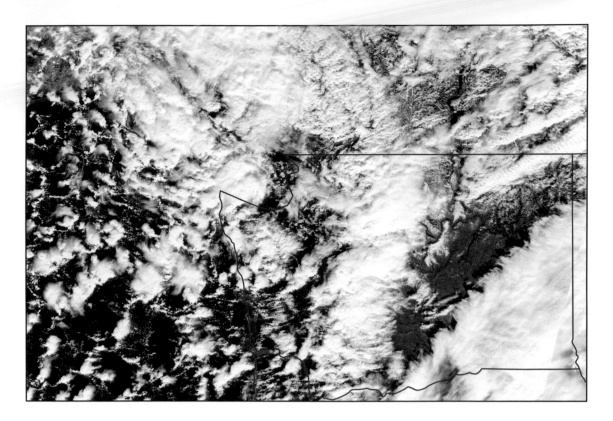

southeastern portions of the Olympics can receive extraordinarily heavy precipitation, sometimes collecting 5–10 inches per day of water and, if temperatures are cool enough, heavy snow can fall over the Kitsap Peninsula to the southeast of the barrier. Such heavy upslope precipitation on the southeastern side of the Olympics can cause Olympic Peninsula rivers such as the Skokomish to overflow their banks, flooding nearby communities.

IS THE NORTHWEST REALLY THAT WET?

The Pacific Northwest has a reputation for being wet and moss-covered, but in reality major Northwest cities receive less annual precipitation than many of their counterparts in the eastern and central United States. As shown in table 2.1, Seattle and Portland receive nearly 20 inches *less* per year than Miami, 10 inches *less* than Houston and

2.8. High-resolution visible satellite picture at 1:00 PM PST on February 20, 2007. As cold showery air approached the Northwest from the west, there was enhancement of clouds on the western slopes of the Olympic mountains, with fewer clouds to their lee over Puget Sound. As air ascends the western side of the Cascades, clouds are greatly enhanced. On the eastern side of the barrier the air sinks and warms, resulting in a rapid lessening of the clouds and virtually clear skies over the eastern slopes of the Cascades. Image from the NASA MODIS satellite.

Atlanta, and about 5 inches *less* than New York and Boston. With extended periods of drizzle and few thunderstorms, Northwest precipitation is typically lighter than the eastern U.S. variety, but the number of days with at least a trace[3] of precipitation is considerably greater over the Northwest than at

---

3   A trace is defined as a precipitation occurrence of less than 0.01 inch of rain, roughly what it takes to make concrete uniformly wet.

LOCATION	ANNUAL AVERAGE PRECIPITATION (IN)	NUMBER OF DAYS WITH A TRACE OR MORE OF PRECIPITATION	NUMBER OF CLOUDY DAYS
Seattle, Washington	38.4	157	228
Portland, Oregon	37.4	151	229
Houston, Texas	46.9	101	166
New York, New York	43.1	120	133
Atlanta, Georgia	49.8	116	146
Miami, Florida	57.1	128	117
Boston, Massachusetts	43.8	126	161

Table 2.1. Long-term averages of precipitation and cloud cover for select U.S. cities, 1971–2000.

locations back east (e.g., 157 days in Seattle versus roughly 120 over eastern states). The Northwest's lead in the number of cloudy days is even more pronounced, nearly 230 per year in Seattle and Portland compared with approximately 160 in Boston and Houston and 117 in Miami. Since Northwest winters are accompanied by many cloudy days and only short periods of (weak) daylight, it is no wonder that wintertime depression (known as seasonal affective disorder, or SAD) hits some residents of the region.

THE NORTHWEST'S "MEDITERRANEAN" PRECIPITATION REGIMES

The Pacific Northwest experiences essentially three weather regimes during the year: a wet season from November through mid-February, a dry mid-summer period from early July through early September, and the transition times of spring and fall. These features and others are illustrated in figure 2.9, which shows the average (1971–2000) daily precipitation for a collection of Washington and Oregon weather stations.

Over Washington State, there is a rapid decrease in daily precipitation from the coast (Quillayute) and Puget Sound (Seattle) to the dry conditions at Yakima, just east of the Cascades (figure 2.9a). Daily precipitation begins to rise again at Spokane as air ascends the western slopes of the Rockies. But even more striking are the large seasonal variations, particularly west of the Cascades. Typically, 50–60 percent of the precipitation over the region's western half falls in November through February, while roughly 75 percent falls in the wet half year from October though March. In contrast, only about 8 percent of the precipitation falls in June through August. In fact, western Washington and Oregon enjoy some of the driest summers in the entire country. East of the Cascades, a similar annual variation occurs, except that the summer months (June through August) include a slightly higher percentage of the annual precipitation (roughly 15 percent) as a result of more frequent thunderstorm activity. This pattern of cloudy, wet winters and dry, sunny summers is known as a Mediterranean climate, since southern Europe experiences a similar variation.

A closer look at Washington's annual precipitation variations reveals interesting details. Although the period from November through February is generally quite wet, the greatest rainfall occurs during the last weeks of November, just in time for Thanksgiving. There is actually a drying trend in December, followed by a plateau of near constant precipitation amounts in January and early February. March begins a steady downward trend in daily rainfall that ends in a period of extraordinary drought during late July and early August—an ideal period to plan a wedding or any outdoor activity.[4] During that magical two-week period, many locations of western Washington receive rain only once in ten years. A careful examination of the daily records reveals a minor increase of showers on the Fourth of July. Some have suggested that the smoke from fireworks contributes to such soggy Independence Days, but no real evidence exists for this theory. The July descent into dry conditions typically begins right after the July Fourth weekend. Thus, the humorous comment within weather circles that summer starts in the Northwest on July 12 is not without some basis. Rainfall increases slowly in September, followed by a steep jump in October. The transition in October is often jarring, going from the generally dry, warm

days of September to the wet, cool, and windy clime of November, often over a period of a week or two. Few regions around the world experience such a rapid turning on of winter.

Over Oregon, precipitation changes throughout the year mimic those of Washington, but with a few differences (figure 2.9b). Again, the coastal zone (North Bend) is far wetter than the western lowlands (Portland). An anomaly is Medford, which although nominally on the western side of the state, experiences far less precipitation than Portland or other west-side locations. Though its more southern position makes a small contribution to Medford's dryness, the key factor is the city's location in a topographic bowl within the Siskiyou/Klamath Mountains, with descending air resulting in drying. Burns, in the central portion of eastern Oregon, receives even less precipitation than Medford. For both Oregon and Washington the geographic variations in precipitation are greatest during the wet season, with only slight spatial differences in precipitation during the midsummer drought.

The large variation in precipitation between summer and winter experienced over the Northwest is quite different from that observed over the eastern two-thirds of the country, where precipitation doesn't vary greatly by season. For example, New York City receives about the same precipitation each month, roughly 4 inches (figure 2.10). In contrast, Seattle's precipitation varies substantially, being far *less* than that of New York for about half the year, and only exceeding the Big Apple from November through February. It is not without some irony that Seattle's new baseball stadium, Safeco Field, was built with a 100 million dollar movable roof, while baseball parks back east,

---

4   The author enjoys holding "dry sky" parties during the drought week at the end of July.

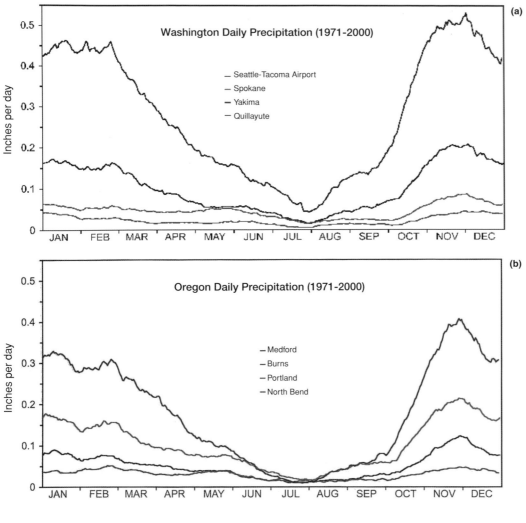

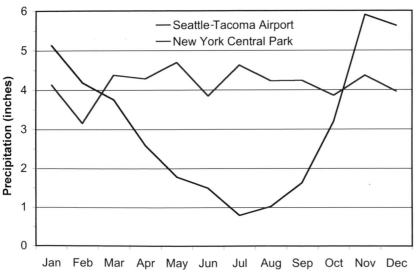

2.9. Daily average precipitation (1971–2000) for four locations in (a) Washington and (b) Oregon. The Northwest gets much of its precipitation during the winter months, with summers being quite dry.

2.10. Monthly mean (1971–2000) precipitation at New York City (Central Park) and Seattle (Seattle-Tacoma Airport). For more than half the year, Seattle is drier than New York.

where summertime rain is far more prevalent, are open to the sky. Stranger yet, Seattle's new football stadium (Qwest Field) does not have a roof, even though the football season coincides with Seattle's wettest period.

Why does the Northwest have such extreme wet and dry seasons? The answer is twofold: the seasonal shifts in the jet stream over the eastern Pacific and our relative lack of summer thunderstorms. During midwinter the Northwest faces an onslaught of weather systems brought on by the intensification and southward movement of the jet stream, the current of strong westerly upper-level winds (figure 2.11). The jet stream, positioned between a strong wintertime Aleutian/ Gulf of Alaska low-pressure area to the north and a weakened east Pacific high-pressure area to the south, serves as a highway for rain-bearing Pacific weather systems. During the summer, the jet stream and associated weather systems weaken and are displaced to the north. In its place, the East Pacific High (or anticyclone) builds northward, resulting in sinking air and drying over the Northwest.[5] Thus a summer regime is established, with dry low-level northerly winds near the surface and far less clouds and precipitation.

Another factor contributing to the Northwest's dry summers is the lack of thunderstorms and convective rain. The Northwest in general and the western side in particular have fewer summertime convective showers and thunderstorms than almost anywhere in the United States. Convection

is the meteorological term for towering cumulus or cumulonimbus (thunderstorm) clouds, with the latter associated with heavy showers and lightning (figure 2.12). The concept of convection is familiar to anyone who has heated water or hot cereal on a stove: as the saucepan warms, a large vertical difference in temperature is created; this results in convection, indicated by rising currents in the water or hot cereal, with descending zones in between. The atmosphere also produces convection when temperature decreases rapidly with height.

During the summer, weak onshore flow off the cool waters of the Pacific (50–55 °F) usually keeps air near the surface relatively cool. Thus, the region west of the Cascades rarely experiences high surface temperatures, which works against the formation of convection. A large supply of water vapor in the lower atmosphere also helps fuel convection and thunderstorms, and ironically the cool temperatures of the Pacific Ocean prevent the air from picking up much moisture, since the amount of water vapor that air can contain increases rapidly with temperature. Furthermore, sinking air aloft associated with a strong east Pacific high-pressure area also contributes to the lack of thunderstorms over western Washington and Oregon. Thunderstorms are somewhat more frequent east of the Cascades, where surface temperatures are higher and occasional "monsoon" moisture from the southwestern states is entrained northward.

Over the eastern and central United States the situation is very different. The summertime Bermuda High, centered over the Atlantic Ocean, pushes air northward over the heated land of the eastern and central states. The air reaching eastern North America during the summer is warm and

---

5   High-pressure areas are generally associated with sinking air that reduces the amount of clouds and precipitation.

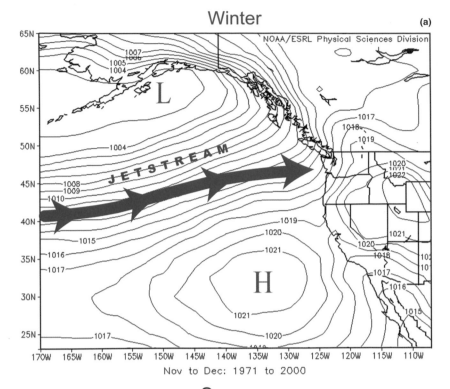

Winter (a)

NOAA/ESRL Physical Sciences Division

J E T S T R E A M

L

H

Nov to Dec: 1971 to 2000

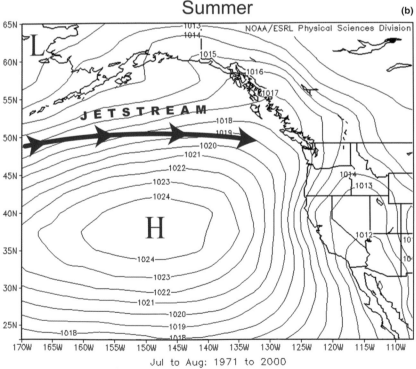

Summer (b)

NOAA/ESRL Physical Sciences Division

L

J E T S T R E A M

H

Jul to Aug: 1971 to 2000

2.11. The differences between (a) winter and (b) summer atmospheric weather patterns help explain the large seasonal variations in precipitation at most Northwest locations. The axis of the upper-level jet stream is indicated by the bold lines with arrows, while the black lines indicate lines of constant pressure. The thickness of the arrows denotes the strength of the jet stream. In winter, the jet stream and the storms associated with it are strong and pass directly over the Northwest, while during summer they weaken and move to the north. High pressure and associated fair weather are pushed far south during the winter, but strengthen and extend northward during the summer. Illustration by Beth Tully/Tully Graphics.

heavily laden with moisture, having passed over the warm Gulf of Mexico, where sea-surface temperatures are often in the lower to mid-80s °F. As this warm, humid air is further heated over land by the sun's warmth, thunderstorms frequently develop east of the Rocky Mountains and produce substantial rainfall. Thus, during the summer the eastern United States gets large amounts of convective precipitation, while the Northwest, one of the most thunderstorm-free regions in the country, gets nearly no precipitation. During the winter, both sides of the continent get precipitation from mid-latitude storms associated with the jet stream.

Why Do Mountains Influence Precipitation?

The key reason for the complex variations of precipitation around the Northwest is the effect of terrain. In discussing the impact of mountains on weather, meteorologists like to use the term "air parcel" to signify an identifiable portion of air—think of a balloon filled with air. As air parcels approach the slopes of a mountain they are forced

2.12. An unusual line of strong thunderstorms along the western slopes of the Cascades. The atmosphere was very unstable that day, with a large change in temperature with height in the lower and middle atmosphere and strong forcing of upward motion by an approaching weather disturbance. Note the beautiful anvil clouds at the top of the convection. Photo courtesy of Frank Jenkins.

to rise (figure 2.13). Since air pressure decreases with height,[6] rising air parcels experience decreasing pressure and expand as they climb the windward slopes of a mountain or hill. This expansion is evident in real balloons, which increase in size as they ascend to regions of lower pressure. Eventually the expansion is so great the balloons burst! Expanding air parcels cool, a fact evident in the coolness of the spray from pressurized aerosol cans or the chilly air escaping from a pressurized tire. It takes energy to expand, and that energy is

---

6   Atmospheric pressure is caused by the weight of the air above. The higher one is, the less air is pressing down from above; thus air pressure decreases with height.

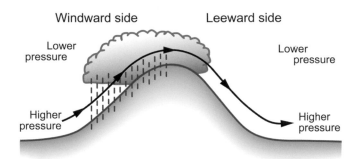

2.13. Schematic of air flow over a mountain barrier. On the windward side, air is forced upward by the mountain and encounters lower pressure as it rises. Lower pressure causes the air to expand and cool. Eventually the air cools sufficiently to become saturated. As it ascends farther, clouds and often precipitation form. On the leeward side the air sinks, compresses, and warms, causing clouds to dissipate. Illustration by Beth Tully/Tully Graphics.

supplied by lowering the air parcel's temperature.[7] But what does decreasing temperature have to do with rain and snow? Just about everything.

All air parcels contain some water vapor and that is particularly true of air approaching the Northwest coast, having traversed thousands of miles of the Pacific Ocean. The amount of water vapor that can be contained in an air parcel depends on temperature, with warmer air able to contain more water vapor than cooler air. Imagine a situation where an air parcel holds a certain amount of water vapor and no clouds or precipitation are apparent. As the air parcel is forced to rise by a mountain, the surrounding pressure decreases and thus the air parcel expands and cools. As the air cools it can hold less and less water vapor. Eventually, if the air parcel rises and cools enough it can only hold the amount of water vapor it already carries and no more. This is called *saturation*, and at this point the relative humidity is 100 percent.[8] If the air parcel rises and cools further, it will have more water vapor than it can hold, so some of the water vapor (a gas) must condense out into liquid form and cloud droplets appear (figure 2.14). Of course, if the temperature is cool enough the condensing water vapor produces clouds of ice crystals instead. If the air parcel continues to ascend the slope, it cools more, resulting in more condensation and an increase in the number and size of cloud droplets or ice crystals. If the cloud droplets or ice crystals grow large enough, they can fall toward the earth, resulting in rain or snow.

In contrast to the windward slopes, air parcels descend on the lee slopes of mountain barriers. As the air parcels descend, the pressure increases, the air parcels are compressed, and their temperature increases. The tendency for warming of compressed air is familiar to anyone who pumps up a tire—the pump housing, where the air is compressed, becomes warm to the touch. As air warms,

---

7   Temperature is related to the speed of atoms and molecules, with warmer temperatures associated with faster motions. The energy of these motions can be used to expand the air parcel, but the result is slower motions and thus lower temperature.

8   Relative humidity is the ratio of the amount of water vapor in an air parcel divided by the maximum possible that it can hold at that temperature. At 100 percent relative humidity, the air parcel can hold no more.

its ability to hold water vapor is enhanced and condensation no longer occurs. In fact, as the descending air parcel warms, water and ice particles tend to evaporate, producing a dramatic reduction in clouds and precipitation. The descent of air over the eastern slopes of the Cascades explains the often rapid decrease of clouds and precipitation east of the Cascade crest when the regional winds are from the west. Figure 2.8, found earlier in the chapter, provides a dramatic example of such mountain effects: the windward side of the Cascades is engulfed in clouds, while the eastern Cascade slopes are essentially clear. Northwest residents who understand such local meteorology can often enjoy a hike or other outdoor activity by taking their recreation to locations with descending air.

2.14. Upslope clouds form as air moves up mountain slopes. This picture shows upslope clouds on high terrain above Kennedy Lake near Tofino, Vancouver Island. Photo courtesy of Michael Hanna.

3

FLOODS

FLOODING DUE TO HEAVY PRECIPITATION AND MELTING SNOW IS THE MOST

serious weather threat for the Pacific Northwest, causing more damage and injury than wind-

storms, snowstorms, or any other meteorological phenomenon. Since 1980, the region has

experienced three weather disasters with damage exceeding one billion dollars, and all were

associated with flooding. In Washington State, twenty-five of the thirty-seven presidential

disaster declarations since 1955 have been for flooding, while flooding in Oregon caused four-

teen of that state's twenty-one presidential disaster declarations. The Northwest's heavy

precipitation, rugged terrain, and steep slopes make it particularly vulnerable to flooding

and to the related threats of landslides and slope failures (figure 3.1).

THE INGREDIENTS OF PACIFIC NORTHWEST FLOODING

Virtually all Pacific Northwest floods are associated with one or more key factors: the influx of warm, moist tropical or subtropical air into the region, rapid snowmelt, or a strong thunderstorm feeding heavy precipitation into a valley or low area. The region's mountains play a major role in enhancing flood threats, even away from their immediate vicinity.

Influx of Tropical or Subtropical Air

The most important ingredient for producing heavy precipitation and flooding over the Northwest is the influx of warm, moist air from the southwest; air that originates in the tropics and subtropics. Often this moist air current is clearly visible in weather-satellite pictures as a band of clouds stretching from near or north of Hawaii to the Northwest coast, and thus is frequently called the *Pineapple Express* (figure 3.2). With an origin far to the south over the warm waters of the subtropical Pacific, Pineapple Express air raises winter temperatures to 10 °F or more above normal, with the western lowlands warming into the 50s °F and occasionally into the mid-60s °F. With such warmth, freezing levels often rise to above 6,000 feet and occasionally as high as 8,000 feet.[1] Since warm air can contain more water vapor than cold air, such warm currents have the potential to bring very heavy rainfall. In fact, up to a few feet of rain can fall in one or two days as warm, moist air is

3.1. Flooding on the Cowlitz River immediately after the heavy rains of the November 5–7, 2006, Pineapple Express event. Photo courtesy of the Lewis County Sheriff's Office and the Lewis County Department of Emergency Management.

3.2. During Pineapple Express events, clouds often extend from near Hawaii northeastward to the Pacific Northwest, as illustrated by this infrared satellite picture taken from 21,000 miles above the surface. This image shows conditions at 9:00 AM PST on November 6, 2006, during a major flooding event over the Northwest. Image from National Oceanic and Atmospheric Administration/National Weather Service GOES-West weather satellite.

---

1   The freezing level is the elevation above sea level at which temperature first drops below freezing.

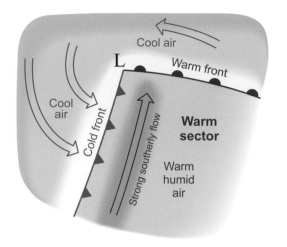

3.3. Maximum flooding potential occurs when warm, humid air within a storm's warm sector is over a region. Strong southerly flow to the east of the cold front can greatly enhance precipitation if it ascends a mountain barrier. The "L" indicates the location of the storm's low-pressure center. Blue and red shades indicate cool and warm air, respectively, and the white shading shows *frontal zones*, where temperature changes rapidly with distance. Illustration by Beth Tully/Tully Graphics.

forced to rise by the mountains, usually in concert with Pacific fronts and storms. As described below, the enhanced warmth of Pineapple Express flows can quickly melt snow, particularly for relatively shallow snowpacks over lower elevations, resulting in an additional source of water.

The potential for Northwest flooding associated with warm subtropical air is greatest when the *warm sector* of a Pacific cyclone (or low-pressure center) passes over the region. To understand this important feature, a little bit of cyclone 101 is in order. Pacific cyclones, the type of storms usually associated with strong winds and heavy rain in the Northwest, generally develop on the boundary between warm air from the tropics and subtropics, and cool air from northern latitudes. In fact, the horizontal change in temperature across this zone

is the main energy source for midlatitude cyclones. The wind field of the developing cyclone distorts this north-south temperature distribution: behind the low center, cold air pushes southward, with its leading edge known as a *cold front*, while east of the low center the storm circulation thrusts warm air northward, with the *warm front* representing the leading edge of the warm air (figure 3.3). The region of warm air between the cold and warm fronts is known as the *warm sector*. The air in the warm sector usually comes from the southwest and, because of its warmth and low latitude origin, contains large amounts of water vapor. Frequently, a strong current of southerly winds occurs in the western portion of the warm sector, just east of the cold front. Flooding potential is greatly enhanced if this southerly current ascends large mountain barriers, such as the Cascades, or if a slow-moving storm allows the warm, moist flow to remain over an area for an extended period, as it did during the flooding of December 3, 2007.

Melting Snow

Melting snow can be a significant contributor to Northwest flooding. A fresh snowfall of 10 inches typically produces about 1 inch of water when melted, and denser or older snowpacks usually contain much more water for the same depth of snow. Several studies have suggested that for the typical Pineapple Express flooding situation, heavy rain is the main cause of flooding, with snowmelt being a secondary factor. However, when snow melts slowly and over an extended period of time, as during the spring, rivers can rise steadily until they reach flood stage. Above-freezing temperatures obviously contribute to snowmelt, but other

3.4. The city of Kennewick during the May–June 1948 flooding. Photo by Robley Johnson and provided courtesy of Richard Johnson.

factors can aid as well, such as strong winds that continually bring fresh warm air in contact with the snow, and warm, humid air that allows condensation on the cold snow surface, producing substantial latent heating.[2]

Prior to the 1930s, when the Columbia River had few dams, snowmelt flooding occurred frequently in the spring, particularly during years with a heavy snowpack and above-normal springtime temperatures. Perhaps the most catastrophic spring flooding of the twentieth century occurred during the

Memorial Day weekend of May 1948 following a period of heavy snow in the mountains of the Columbia River basin. After a few weeks of above-normal temperatures and heavy rain during the later part of the month, the Columbia River rose significantly, inundating Richland and Kennewick in eastern Washington on May 31. With only two dams in place (Grand Coulee and Rock Island), the ability to store water for later release was limited. In Kennewick, several feet of water covered an area approximately 2 miles long and a half mile wide, flooding many local businesses (figure 3.4) and resulting in one death. Fortunately, the downtown area of Richland

---

2   It takes energy or heat to evaporate water. That energy is not lost, but is reclaimed later when the water vapor, a gas, condenses back into liquid water. Such warming, associated with the condensation of water vapor into liquid water, is known as latent heating, since the warmth is hidden, or latent, until condensation occurs.

was saved by the rushed construction of a twelve-foot dike around the city.

Although flood damage in the Tri-Cities totaled nearly fifty million dollars, the greatest impact of the rising Columbia River occurred downstream near present-day Portland, where the town of Vanport was destroyed by floodwaters. Vanport was built in the early 1940s to house the thousands of shipyard workers that flocked to Portland during the war. Approximately eighteen thousand people were living in Vanport's sprawling collection of wood-frame apartments on Sunday, May 30, when the rising river broke through the dikes surrounding the city and quickly inundated it with over 15 feet of water. Without any warning of the flooding, twenty-five residents lost their lives and all were deprived of their homes and possessions. The economic loss exceeded one hundred million (1948) dollars, and the town of Vanport no longer exists.

3.5. Daily precipitation (inches) at Salem, Oregon, at an elevation of 210 feet, and at Little Meadows, a site on the slopes of the Cascade Mountains at 3,999 feet, during a relatively wet period in November and December 2001. On most days, high-elevation Little Meadows received at least twice as much precipitation as Salem.

Although melting snow has contributed to past floods, snow can also reduce or delay flooding. Snowpacks, and particularly deep snowpacks, have the ability to soak up large amounts of rain, which can then refreeze within the snow. This water is held in the snow until later in the season and thus can mitigate the potential for flooding during a Pineapple Express or other rain-on-snow event. This spongelike quality of snow allows the amount of water in the snow, called the *snow water equivalent*, to increase during the winter, even when the snow depth remains the same or decreases.

Terrain Enhancement of Precipitation

As discussed in the second chapter, clouds and precipitation are greatly enhanced when air is forced to ascend the windward slopes of mountain barriers. Most major Northwest flooding events start with an extensive region of light to moderate precipitation linked to a strong Pacific low-pressure system and its associated fronts. This precipitation is then greatly increased, sometimes by factors of two to five times, as air ascends the mountains. Conversely, precipitation can be less-

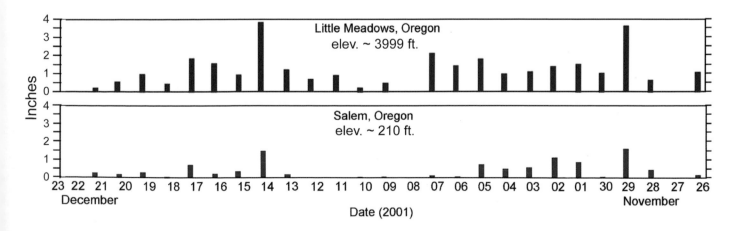

3.6. Aerial view of the terrain surrounding Lake Whatcom, looking toward the south. The city of Bellingham is in the lower right corner, and the lowlands of Puget Sound and Mount Rainier are in the upper right. Graphic generated using Google Earth Pro, printed courtesy of Google Inc.

ened when air moves down mountains, as in the rain shadows to the lee of the Olympic and Cascade mountains. The profound influence of mountains on precipitation is illustrated in figure 3.5, which shows precipitation during a wet, late-fall period in 2001 at Salem, Oregon, a lowland station (210 feet above sea level) within the center of the Willamette Valley, and at Little Meadows, a station high (3,999 feet) on the nearby western slopes of the Cascades. Precipitation amounts were greater at Little Meadows on all days but one, with the mountain location generally receiving at least twice as much precipitation as Salem. Such heavy precipitation on steep mountain slopes quickly finds its way into streams and rivers, resulting in rapid increases in both stream flow and river heights.

Although the heaviest precipitation tends to fall on the middle and upper windward slopes of mountain barriers, under certain conditions heavy rain or snow can fall on the lower slopes where the terrain begins to rise abruptly. This can happen when air is close to becoming buoyant, so that a modest upward push of air parcels by rising terrain results in the air rushing upward, forming cumulus clouds and thunderstorms.[3] Many a Seattle skier has witnessed this effect while driving eastward toward Snoqualmie Pass on Interstate 90, with heavy showers forming in the vicinity of the foothills city of North Bend, Washington.

A dramatic example of an increase of precipitation caused by mountain foothills occurred during January 9–10, 1983. After an extended period of above-normal rainfall, a Pacific low-pressure system moved eastward across southern British Columbia. A warm front crossed the region, resulting in warm, moist air flooding western Washington and the freezing level rising to about 6,000 feet. The strong southerly flow approached the low mountains (with peaks around 3,000 feet) surrounding Lake Whatcom, which is located south and east of Bellingham (figure 3.6). The

---

3 Meteorologists have a term for air that will become upwardly buoyant after a sufficiently large upward push: conditionally unstable. Such air is usually associated with a large change in temperature with height and high levels of water vapor at low levels. A buoyant air parcel is analogous to a helium balloon that accelerates upward when released.

3.7. In January 1983 after localized heavy rain in the mountains above Lake Whatcom, water and debris descended Smith Creek, causing extensive damage to lakefront homes. Photo courtesy of Professor Don Easterbrook, Western Washington University.

incoming flow became upwardly buoyant as it rose on the mountain slopes, resulting in intense rain showers over the Lake Whatcom area. A number of locations near the lake and its immediate surroundings reported 6–8 inches of rain over the twenty-four hours ending the morning of January 10, while Bellingham Airport, a few miles away, collected only 1.1 inches. Even reporting stations high in the Cascades only received 3–5 inches for the entire storm. Heavy localized rain over the hills surrounding Lake Whatcom, and the rapid melting of some of the 1–2 feet of snow on the mountain slopes, resulted in an extraordinary quantity of water reaching the already saturated slopes. Water

surged down streambeds and mixed with dirt, logs, and other debris to produce fast-moving debris flows that descended several creeks on the western side of Lake Whatcom. The debris flows and storm flooding destroyed twenty-six homes and extensively damaged thirty-five, with losses exceeding ten million (1983) dollars (figure 3.7). At the same time, the Iowa Street portion of the Bellingham commercial district was flooded, a large portion of the nearby Sudden Valley Golf Course was destroyed, twelve train cars were derailed south of Bellingham, and extensive flooding hit farms along the Nooksack River.

Summertime Convective Storms

Although strong convective events such as thunderstorms are less frequent over the Northwest than over the eastern half of the United States, they can produce severe localized flooding and slope

failures (such as mudslides) under the right conditions. The first ingredient is the occurrence of a strong thunderstorm capable of producing heavy rain. Such thunderstorms are most frequent during the late spring and summer east of the Cascade crest, where the marine influence is reduced and temperatures are warmer. Slow-moving or stationary thunderstorms, often forced by terrain features, are the most damaging, since they allow large amounts of water to accumulate in one area. Relatively narrow valleys or watersheds where rain can be concentrated are also major contributors to such events. Flooding associated with summertime convective events can build rapidly and with little warning, since thunderstorms are often poorly forecast and can develop quickly. Thus, the term *flash flooding* is often used to describe these events.

One of the most well-known Northwest flash-flood episodes occurred in the northeast Oregon town of Heppner on June 14, 1903. Heppner, located about 10 miles northwest of the northern slopes of the Blue Mountains, was founded within a narrow valley drained by Willow Creek and its tributaries. By the middle of the afternoon that day, thunderstorms began to build to the southwest of town, and by 4:30 PM heavy rain and hail started to fall in town. In a half hour, the streams began to flood and a fast-moving wall of water, roughly 5 feet high, pushed through town. Two-thirds of the town's homes were destroyed, the business district was leveled, and nearly 250 lost their lives (figure 3.8). More recently, on July 3, 1998, a flash flood struck the Yakima River canyon in the area north of Yakima, Washington. During that afternoon, thunderstorms developed along the crest and eastern slopes of the Cascades, producing 3 to 4 inches of rain in less than an hour over the Yakima drainage basin. The heavy rain triggered massive mudslides along the slopes of the canyon that covered eight sections of scenic State Route 821 to depths exceeding 15 feet, trapping motorists, campers, and a large freight train. The heavy rain also caused scattered power outages and many streets and intersections in Yakima were flooded.

Willamette River 1996 Flood

3.9. Heavy precipitation over northern Oregon produced extensive flooding of the Willamette River valley during February 1996. Photo courtesy of the National Weather Service, Portland.

MAJOR RECENT PACIFIC NORTHWEST FLOODS

February 5–9, 1996

The heavy rain and flooding of February 5–9, 1996, was one of the most damaging Northwest weather events on record, producing approximately a billion dollars of damage and the loss of nine lives in Oregon, Washington, Idaho, and western Montana. More than ten thousand buildings and homes were damaged or destroyed, displacing over thirty thousand residents and hundreds of businesses. Rivers throughout western Oregon and Washington experienced major flooding, with some achieving record levels (figure 3.9). With flooding and hundreds of mudslides and landslides, roads were closed throughout the region, including key routes such as Interstate 5, Interstate 90, and US 2. The February 1996 storm was a classic Pineapple Express event with heavy rain, warm temperatures, and considerable snowmelt.

Prior to February 1996, the winter had been unusually wet, with many locations receiving 125–175 percent of normal precipitation, leaving the soils saturated and filling streams and rivers. Although the mountain snowpack was below nor-

mal at mid-January, a subsequent transition to a much colder regime resulted in heavy snow and the development of above-normal mountain snow depths by the end of the month. Significantly, even the lower slopes of the Cascades gained substantial snow. By February 6, the atmospheric circulation shifted so that strong southwesterly flow originating in the tropics headed directly toward the Northwest, resulting in heavy precipitation and high freezing levels of 6,000 to 8,000 feet.[4] Embedded in this warm, moist flow were a series of weather disturbances that produced extraordinary rainfall totals over a four-day period, particularly over the windward mountain slopes. During February 5–9, rainfall totals ranged as high as *25–30 inches* over the coastal mountains as well as the Cascades of northern Oregon and southern Washington, with even the Willamette Valley and

---

4   As noted earlier in this chapter, the freezing level is the height above the surface at which the temperature falls below freezing. Wet snow can fall as much as 1,000 feet below the freezing level. Normal freezing levels along the western slopes of the Cascades during the winter are around 3,000 feet.

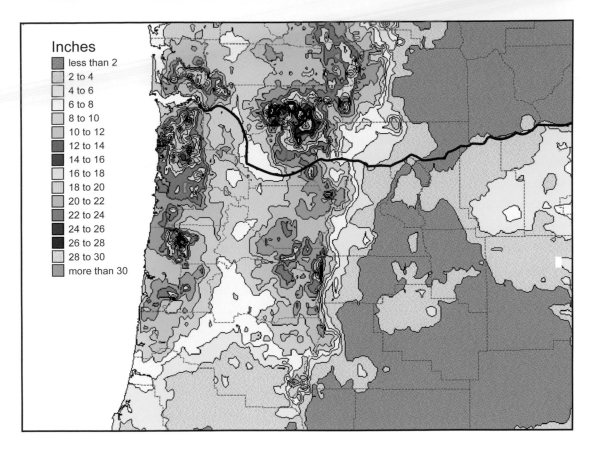

Inches
- less than 2
- 2 to 4
- 4 to 6
- 6 to 8
- 8 to 10
- 10 to 12
- 12 to 14
- 14 to 16
- 16 to 18
- 18 to 20
- 20 to 22
- 22 to 24
- 24 to 26
- 26 to 28
- 28 to 30
- more than 30

3.10. Four-day precipitation totals over northwest Oregon and southwest Washington during the February 5–9, 1996, flooding event. Graphic courtesy of Chris Daly and Mike Halbleib of the Oregon State University PRISM group.

western Washington lowlands getting 7–10 inches of rain (figure 3.10). For example, Laurel Mountain (3,600 feet), in the Oregon coastal range east of Salem, received nearly 30 inches during those four days; Astoria collected 8.88 inches; and the Portland Airport was soaked by 7 inches. At a number of locations, the four-day precipitation totals represented all-time records for such a period.

This heavy rain, coupled with the rapid melting of the low-elevation mountain snowpack, resulted in extraordinary quantities of water reaching the region's streams and rivers, producing some of the worst flooding in sixty years.[5] Flooding occurred in over twenty-one rivers of northwestern Oregon;

six experienced all-time record floods, and the Columbia and Willamette rivers rose 10–20 feet above flood stage. Nine major western Washington rivers overtopped their banks, and flooding even occurred on rivers east of the Cascades, such as the Yakima, which reached its second-highest level of all time. Interstate 5, the region's main north-south transportation corridor, was closed due to 6–8 feet of water over the roadway at Chehalis, Washington, and landslides closed both Interstate 90 and US 2, preventing east-west travel across Washington. Thousands of landslides and slope failures occurred over the region, with the greatest number on the northern slopes of the Blue Mountains (in

---

5   Combined rainfall and snowmelt provided more than 20 inches of total water over many mountainous areas, with a few windward slopes receiving nearly 40 inches of total water equivalent (combination of rain and melted snow).

the southeast corner of Washington) and to the south of Mount Saint Helens, where precipitation was heaviest. In Seattle and environs, hundreds of landslides occurred on steep hill slopes, damaging or destroying several homes. Around the region, rail traffic was halted as slides covered tracks and even contributed to the derailment of several freight cars. As a result of the regional flooding and landslides, approximately sixty counties in Oregon, Washington, Idaho, and Montana were declared federal disaster areas.

Although Pineapple Express events occur regularly over the Pacific Northwest, the February 5–9, 1996, rainstorm was extraordinary due to the unusual several-day duration of the heavy rain as well as the exceptionally warm temperatures: the daily low temperatures at many stations were in the 50s °F, which is considerably higher than the normal high temperatures during that time of the year. The heavy snowpack, particularly at lower elevations, provided massive amounts of meltwater during the extended warmth. Another unusual aspect of the February 1996 event was that heavy rain and melting snows extended into Idaho and Montana. In Idaho, a number of locations experienced five-day precipitation totals of 4–8 inches and, with the rapid melting of recent snows, flooding occurred around hundreds of streams and rivers. Damage in that state was in the tens of millions of dollars, hundreds of homes were destroyed, and one individual lost his life.

Late December 1996 to mid-January 1997

The period of late December 1996 through mid-January 1997 was the "perfect storm" of Northwest flooding and landslides because it was initiated by a combination of unusual events. The year 1996 was the third wettest of the previous century, and December was particularly soggy: many Northwest locations received twice their normal monthly precipitation, with particularly heavy rains over the southern portions of Oregon, where *three times* the normal precipitation was observed. The region's soils were saturated, and groundwater and stream and river levels were high. During the last week of December a cool and persistent flow from the Gulf of Alaska produced extensive and deep low-elevation snow over much of the Northwest. Seattle-Tacoma Airport received 13 inches of snow on December 28–29, the second-largest 24-hour amount on record, with 19 to 23 inches of snow on the ground from north Seattle into Snohomish County. Eastern Washington was also covered by extraordinary snowfall, attaining an all-time record depth of 27 inches at Yakima on December 29. On that day, the winds aloft became southwesterly and a very warm, moist flow became established. Temperatures at Seattle-Tacoma Airport rose from 27 to 49 °F in a few hours, accompanied by heavy rain and strong winds gusting to 49 miles per hour.

This combination of warmth, rain, and strong winds produced rapid melting of lowland snows. The collective effect of heavy rain and melted snow brought extraordinary amounts of water to the already saturated soils. In fact, for the central Puget Sound basin no previous storm on record had ever hit the ground with so much water in so little time. Street drains throughout the region became clogged with slush and debris, as did the gutters of innumerable local residences. Hundreds of streets were flooded, and saturated steep hillsides began to fail, particularly along the bluffs overlooking Puget Sound and Lake Washington (figure 3.11). The heavy rains and warm temperatures continued for the next week, contributing to hundreds of slope failures and landslides that destroyed dozens of homes and roads.

Some of these slides occurred weeks after the heavy rains as the water percolated into the soil. For example, during the early morning of January 15, 1997, a large landslide occurred on the Puget Sound bluffs south of Edmonds in the town of Woodway. The failed hillside slid into the Sound, knocking five freight cars of a passing train into the water. Four days later, on January 19, 1997, tons of rock, trees, and soil crushed a home beneath a bluff on Bainbridge Island, killing all four family members (figure 4.5). In the southwestern Oregon city of Ashland, well-known for its annual Shakespeare Festival, the combination of saturated soils, heavy precipitation, and snowmelt on nearby mountains resulted in a rapid rise of Ashland Creek, which runs through the city. On January 1, 1997, the creek overflowed into downtown Ashland and severely damaged businesses and Lithia Park, resulting in millions of dollars of destruction (figure 3.12).

November 2006

November 2006 brought record-breaking rainfall and flooding to much of the Pacific Northwest. By the end of the month the majority of Washington and Oregon had received 200 percent or more of normal precipitation, many November or all-time monthly records were shattered, and devastating flooding and slope failures closed roads and destroyed structures throughout the region. Mount

3.11. The combination of saturated soils, rapidly melting snow, and heavy rain resulted in the failures of many steep slopes along bluffs and hillsides throughout the Puget Sound region during late December 1996 and the early part of the following month. Photo courtesy of the *Seattle Times*.

3.12. Ashland, Oregon, on January 1, 1997. Ashland Creek overflowed into downtown Ashland after a period of extraordinarily heavy rains. Photo courtesy of Jim Craven, *Medford Mail Tribune.*

Rainier National Park was particularly hard hit, with the greatest storm-related damage in its century-long history. By the end of the month, presidential disaster declarations were made for eleven counties in Washington and four in Oregon, with regional damage totals in the hundreds of millions of dollars.

During that November a progression of approximately a dozen significant Pacific frontal systems struck the Northwest. Of these storms, one stood alone in terms of its duration and severity: the deluge that soaked the region on November 5–7. This event had all the characteristics of a major Pineapple Express downpour: a "river" of moisture extending from just north of Hawaii toward the Northwest accompanied by warm air that drove freezing levels to above 10,000 feet (see figure 3.2 earlier in this chapter). During that three-day period, 4–10 inches of rain fell over the lowlands of Oregon and Washington, with torrents of 10–40 inches in the Cascades, coastal mountains, and the Olympics.

At the Paradise Ranger Station on Mount Rainier nearly 18 inches of rain fell in thirty-six hours on ground that was already saturated from previous storms. The resulting damage was unequalled in park history: dozens of roads were undermined or destroyed; the Sunshine Point Campground was nearly washed away; trails, pathways, and bridges were destroyed; and the entire park was forced to close to traffic for months (figure 3.13). Near Mount Rainier, a 20-year-old elk hunter from Seattle was swept into the Cowlitz River after the riverbank collapsed beneath his pickup truck.

Heavy precipitation forced most western

Washington and Oregon rivers to flood stage, with twelve reaching their all-time record levels. Flooding spread across the region, inundating farms and roads. The Snohomish River breached its protective levees, causing local flooding and closing State Route 9, a major regional highway. In Oregon, landslides closed State Route 35 near Mount Hood, while farms, houses, and businesses were flooded around Tillamook on the Oregon coast. Lee's Camp in Oregon's coastal range received 14.3 inches in a single day, probably representing the wettest day in Oregon's historical record for any location.

Many observing sites in Oregon and Washington beat their all-time November precipitation records, while some exceeded their records for any month.

3.13. The torrential rains and floods of November 5–7, 2006, produced unprecedented damage throughout Mount Rainier National Park. Many roads, such as State Route 123 at milepost 10.5 shown here, were covered with debris, washed away, or undermined. Photo courtesy of Mount Rainier National Park.

Seattle-Tacoma Airport received 15.63 inches, the greatest monthly total since record keeping began there in 1948. Although many newspapers claimed this was the wettest month in Seattle history, the record probably belongs to December 1933, when Seattle observations were taken at the Federal Building downtown. Hoquiam and Stampede Pass also beat their all-time monthly precipitation records, while Vancouver and Olympia exceeded their previous November records. Twelve precipitation sites in Oregon set all-time monthly records, topped by the extraordinary 49.59 inches at Laurel Mountain in the coastal range, which beat the previous (December 1996) record by roughly 12 inches. Another thirteen Oregon locations, including Portland and Astoria, exceeded their previous November records.

December 2–3, 2007

The simultaneous occurrence of sustained hurricane-force winds and heavy rain associated with a strong Pineapple Express is quite rare west of the Cascades, but such an event occurred during December 2–3, 2007, resulting in extraordinary rainfall totals over western Washington and devastating winds over the coastal zone (a description of the winds is given in chapter 5). Twenty-four-hour rainfall amounts of 3–5 inches were typical, with 8–12 inches over Kitsap County, the windward (southern and southwestern) sides of the Olympics, and the Willapa Hills of southwest Washington. Many locations received all-time record rainfall for a twenty-four-hour period, exceeding the one-day totals of the February 1996 event. Some sites in the Willapa Hills collected 14–18 inches of rain over the two-day event. The Chehalis River and its tributaries were particularly hard hit, with some stream gauges

(a)

(b)

3.14. After nearly a day of heavy rain over southwestern Washington, the Chehalis River broke through protecting levees on December 3, 2007, and inundated its floodplain, causing extensive damage to homes and businesses. A particular problem was the flooding of Interstate 5, cutting the key north-south transportation route for the region: (a) aerial view of the flooded interstate and surrounding land near Centralia; (b) ground view of a stretch of the water-covered interstate. Photos courtesy of the Washington State Department of Transportation.

indicating flows double that of previous records. Flooding was extensive around Chehalis and Centralia, inundating homes, farms, and businesses, with the flooded Interstate 5 closed for several days (figure 3.14).

IMPACT OF HUMAN ACTIVITIES ON PACIFIC NORTHWEST FLOODING

Effects on Regional Rivers

Although natural causes, such as heavy rain or rapid snowmelt, are the major factors behind Northwest flooding events, human alterations of the landscape have modified their impact. On one hand, the hydroelectric dams along the Columbia River, with their ability to store much of the spring runoff from melting snow, have greatly reduced the springtime floods that were nearly annual events. On the other hand, some rivers have become more flood-prone as silt and debris, produced by logging and other human activities, have accumulated in river channels. An extreme example of this effect is the Skokomish River, which drains the southeastern side of the Olympic Mountains. Of all western Washington rivers, the Skokomish is generally the first to flood and the one that floods most often (figure 3.15). Until about 1950, the Skokomish acted like other Northwest rivers, topping its banks only during periods of unusually heavy precipitation, but after about 1960 it began to flood more frequently. Research by Professor David Montgomery and his student Cheryl Stover at the University of Washington suggested the cause was twofold: heavy logging along the south fork of the river generated erosion-produced sediments that filled the channel, and the construction of two dams on the north fork of the river reduced river flow, making the north fork less able to scour out sediments. In other locations, debris left in streams during logging produced dams that have broken free during heavy precipitation with terrible effects downstream, as described in the earlier discussion of

3.15. The Skokomish River, draining the southeast side of the Olympic Mountains, is the most flood-prone river of the region due to the buildup of sediments in its channel. Nearby roads, such as the one shown above, are often flooded, with unfortunate effects for the river's salmon population. Photo courtesy of the *Seattle Times*.

the debris flows in the hills above Lake Whatcom. Although mudslides and slope failures have occurred for millennia, both human disturbance of natural vegetation and construction activities (e.g., logging roads, clear-cutting) have greatly increased the frequency of these events.

Another issue is development on river floodplains. Since rivers naturally flood adjacent low-lying areas, development in such locations endangers both life and property. Rivers also meander and shift course in time, further enhancing the threat in their neighborhoods. Building levees to protect areas from the river floods often worsens flooding at other locations and can produce sudden and catastrophic flooding when the levees fail. During the December 2007 floods in Centralia/Chehalis, extensive damage occurred to new commercial developments in the river floodplain, with islands of elevated fill for the new buildings worsening the flooding for neighboring, but lower, homes and businesses.

A major unknown today is whether global climate change due to human-caused emissions of greenhouse gases will affect the region's heavy precipitation and flooding events. Described in chapter 12, current research has not provided conclusive evidence that global warming will result in a greater occurrence of heavy precipitation or flooding.

Urban Flooding

Although rural flooding along the floodplains of major Northwest rivers is frequently in the news, serious flooding can also occur within the urban cores of Northwest cities when unusually intense rainfall occurs. For a number of reasons, Northwest urban areas are particularly susceptible to localized flooding. First, urban areas are covered with roads and other impervious surfaces that lessen the ability of the landscape to absorb or drain water. Second, drains and drainage systems in most Northwest cities are designed to handle the region's usually frequent but relatively light rainfall, in which only a few hundredths of an inch of rain typically falls per hour. In contrast, over the eastern and central United States, strong thunderstorms and intense rainfall are more frequent—ranging from half an inch to several inches an hour—and municipal drainage systems are designed with such storms in mind. Third, heavy Northwest rains often occur in late November and early December, a period when fallen leaves cover the landscape and often clog street and other drains. Also at this time, the ground has had suf-

ficient time to become saturated by the autumn rains. Finally, Northwest urban areas are often hilly, so that rainfall can collect in and flood low-lying areas.

A recent example of urban flooding occurred on December 14, 2006, during the afternoon preceding the Hanukkah Eve windstorm (see chapter 4). As noted above, the previous month had been extraordinarily wet and thus the soils were already saturated. During the afternoon of December 14, steady moderate rain fell as the warm front of the advancing storm swept over the region; many western Washington locations received one or two inches of rain in the six hours preceding 4:00 PM. But it was during the next hour that catastrophe would strike.

A Pacific front, separating relatively weak southeasterly winds from a much stronger flow from the south to southwest, moved inland across the Puget Sound basin between 4:00 and 5:00 PM. The confluence of these two air currents enhanced upward motion[6] and the development of strong convection, thunderstorms, and intense rainfall. As shown in figure 3.16, from the National Weather Service Camano Island radar at 4:44 PM, the front and associated very heavy rain (red shading) were oriented from southwest to northeast, a direction nearly parallel to the wind in the lower atmosphere. Thus, instead of moving through rapidly, as with most isolated convective showers, the heavy rain continued for roughly a half hour as the line slid northeastward over the central portion of Seattle and parts of Kirkland. The result:

---

6   When two low-level air streams converge toward each other, air is forced upward, often resulting in clouds and precipitation.

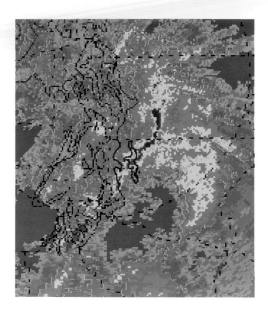

3.16. Image from the National Weather Service radar on Camano Island for 4:44 PM on December 14, 2006. Radar intensity is indicated by color, with the red shade associated with the extraordinarily heavy rain stretching southwest-northeast across Seattle and extending into Kirkland. Yellow indicates heavy rain, and green signifies moderate intensity. Seattle and Lake Washington are outlined in blue.

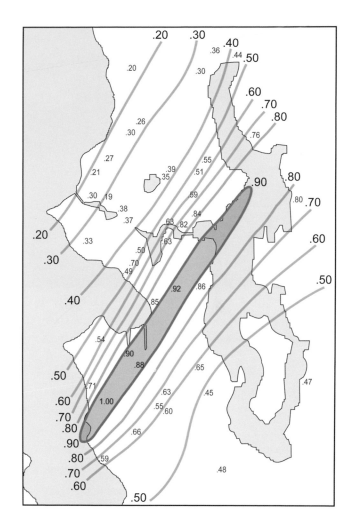

3.17. Maximum one-hour precipitation (in inches) during the late afternoon of December 14, 2006. For most observing sites in the flood area, the most intense rainfall was between 4:00 and 5:00 PM. Note the swath of heavy rain, approaching 1 inch in an hour, running from West Seattle through Kirkland. Red numbers are the one-hour rainfall totals. Illustration by Beth Tully/Tully Graphics.

extraordinary one-hour rainfall totals, peaking at nearly 1 inch per hour over the central portion of the rainfall swath, while 10 miles to the north and south less than half as much fell (figure 3.17). At the University of Washington, where 0.63 inch fell over an hour, nearly half (0.30 inch) of that rainfall occurred in nine minutes, with one sodden minute receiving 0.05 inch. A statistical analysis of this event using long-term historical records indicated that the half-hour rain totals associated with the band of heavy rain were of an intensity only expected to occur once every fifty to one hundred years.

The effects of the storm were stunning. In a swath from West Seattle to the Madison Valley and then through the University District and Kirkland, intense rainfall filled roads to curb top, and parking lots, like those in Seattle's University Village, became lakes. Dozens of major roads became impassable due to flooding and mudslides, result-

3.18. During the evening of December 14, 2006, intense local rain caused extensive urban flooding over portions of Seattle and neighboring communities. Several feet of water collected on Seattle's Mercer Street under the Aurora Avenue overpass, resulting in stalled and floating cars. Photo courtesy of Ricky Fischer.

ing in region-wide gridlock. Cars were afloat on Mercer Street under the Aurora Avenue overpass (figure 3.18), Interstate 5 was closed temporarily due to deep water covering the road, and thousands of people en route to a midweek Seattle Seahawks football game were stranded in their cars for hours. Hundreds of homes were flooded around the city, as sewage and wastewater surged out of drains and toilets. The greatest tragedy occurred in Seattle's Madison Valley, where a woman was trapped in a windowless basement room and was drowned by rapidly rising waters.

Urban flooding returned to much of the Seattle metropolitan area during December 2–3, 2007. As noted earlier in this chapter, these dates brought extraordinary rainfall to the lowlands and coastal regions of the Northwest. Over the western and

northern portions of Seattle, all-time record twenty-four-hour rainfall amounts of 4.5–5.5 inches were observed, representing the kind of event that would be expected to occur once in a century. Roads throughout Seattle and its suburbs flooded, large sinkholes appeared, and landslides occurred over steep slopes. A prime example was the collapse of Golden Gardens Drive NW, a primary route to Puget Sound and the Shilshole Bay Marina (figure 3.19). A major difference between the 2006 and 2007 urban flooding events was that the 2006 rainfall was highly localized and short-lived, with extreme rainfall falling for less than an hour. In contrast, the December 2007 rains were not as intense, but were more extensive geographically and lasted for eighteen to twenty-four hours. A mitigating factor for the December 2007 rainstorm was that the previous month was nearly two inches drier than normal; if the ground had been saturated, as it had been for the December 2006 event, the flooding and destruction would have been far worse.

Recently, the regional press has noted that several "one-hundred-year" rainstorms have buffeted the region, with a hint that this might be the work of global warming. Others have suggested cynically that some sort of statistical chicanery is at foot. It is true that several major rain events have recently struck western Washington, but they have not occurred at the same location. Basic statistical concepts indicate that the more locations one monitors, the larger the chance of getting a record at one of them. Consider getting heads three times in a row in a coin toss. With a fair coin, that should happen once in every eight attempts. But if you ran the experiment simultaneously with eight sets of coins, you would greatly increase your chances

3.19. Record twenty-four-hour precipitation struck north Seattle during December 2–3, 2007, producing extensive flooding in low-lying areas, landslides, and slope failures. As shown in this picture, heavy rainfall on the bluffs overlooking Puget Sound resulted in the collapse of a portion of Golden Gardens Drive NW.

of securing such a string of heads. Furthermore, there is the potential for setting records for various durations of rainfall: 1 hour, 6 hours, 12 hours, and 24 hours are a few examples. Thus, there are even more ways to set a record, since it is possible to set a record for a short-duration rainfall but not a longer one, and vice versa. David Hartley, a hydrologist at Northwest Hydraulic Consultants in Seattle, has used the twenty-nine-year record of seventeen rain gauges in Seattle to calculate that

one would *expect* a "hundred-year" rainfall for some time duration (6, 12, or 24 hours) every three years at one of the gauges. As discussed in chapter 12, starting with basic atmospheric principles one could argue for either increasing or decreasing heavy rainfall over the Northwest in the future; furthermore, current-generation high-resolution climate models do not suggest that global warming will necessarily bring more intense rainfall to the region.

4

SNOWSTORMS AND ICE STORMS

|||||||||||||||||||||||||||||||||||||||||||||||||||||||||||||||||||||||||||||||||||||||||||||||||||||||||||||||||||||||||||||||||||||||||||||||

PACIFIC NORTHWEST SNOW IS A STUDY IN CONTRASTS: WHILE THE WESTERN SLOPES

of the Cascades can receive several feet in a day and nearly 100 feet in a year, the western

lowlands are snow-free for most of the winter, with even a few inches garnering the rapt, if not

excessive, attention of the local media. During the winter of 1998–99, the Mount Baker Ski Area

at 4,200 feet in the North Cascades received 1,140 inches (95 ft), the world record for the larg-

est verifiable snowfall,[1] while during the same winter only a few inches fell in the Puget Sound

lowlands. This chapter explains why snow can be plentiful in the mountains and sparse over

the lower elevations of western Oregon and Washington, and it describes the rare combination

---

1   The old record was 1,122 inches set at Paradise Ranger Station on Mount Rainier during the winter of 1971-72.

of events that make lowland snow possible. It also examines the dangers of a less frequent meteorological cousin: ice storms. Such events often occur within the Columbia River gorge as well as in the Cascade mountain passes and the lower elevations of eastern Washington.

SNOW AMOUNTS OVER THE PACIFIC NORTHWEST

The lowlands of western Oregon and Washington generally do not receive heavy snow: average annual snowfall ranges from approximately 5 inches over the southern Willamette Valley to nearly 15 inches near Bellingham (figure 4.1). At these lower elevations, snow is normally limited to the period from mid-November through the first week of March, with the most snow in December and January. Major snowstorms, with snowfalls reaching over a foot, are rare in the western lowlands, but can occur with the unusual combination of frigid air and moisture.

The mountains of the Northwest are a different story since the higher elevations are some of the snowiest locations in the United States. Much of the terrain above 3,000 feet receives several hundred inches of snow per year, with some locations on the windward slopes (such as the Mount Baker ski area and Mount Rainier's Paradise Ranger Station) buried by 500–700 inches in a typical year and over 1,000 inches during exceptional ones. On the eastern side of the Cascades, a region that is considerably drier and colder during the winter, a typical winter brings only 5–10 inches of snow in the lower Columbia Basin and as much as 100–200 inches over the higher terrain of eastern Oregon and Washington.

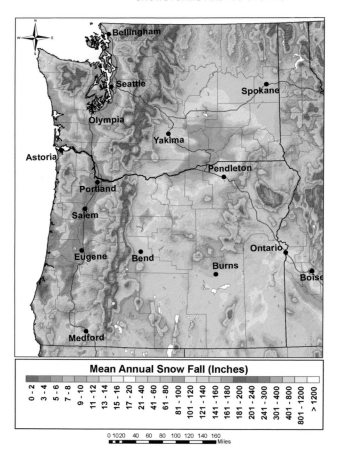

4.1. Average annual snowfall totals (inches) over the Northwest for the period 1960–90. Graphic courtesy of Chris Daly and Mike Halbleib of the Oregon State University PRISM group.

A characteristic of Northwest lowland snow is its substantial year-to-year and decade-to-decade variability, with a trend toward less snow since the mid-1970s. For example, at McMillin Reservoir, 12 miles southeast of downtown Tacoma and at an elevation of 579 feet, heavy snow years were more frequent from the late 1940s though the mid-1970s than thereafter, with the biggest snowfall, 42 inches, occurring during the cold winter of 1949–50 (figure 4.2). The 115-year record at Spokane (2,400 ft) shows similar variations, with large year-to-

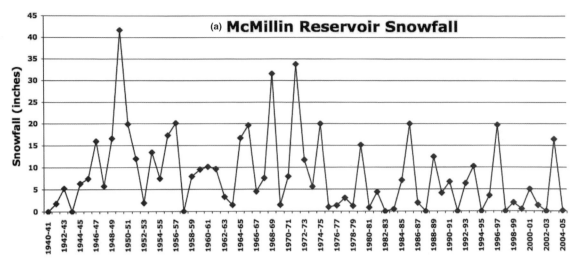

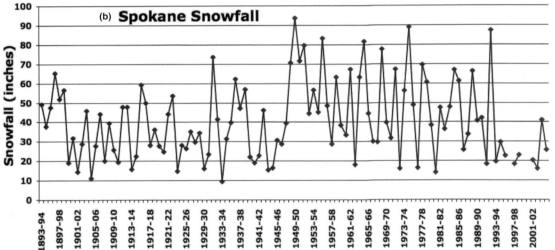

year swings, a maximum in 1949–50, and a relative dearth of snow during recent years.

There are several potential explanations for the decline in lowland snow since the 1970s over the Pacific Northwest. Observations suggest that there has been warming in winter temperatures over the region, averaging about 1–2 °F since the mid-twentieth century. Although such warming may seem small, it could have an impact over the lowlands of western Oregon and Washington, where temperatures are often borderline for snow. There are sev-

4.2. Annual snowfall (inches) at (a) McMillin Reservoir near Tacoma, and (b) Spokane.

eral possibilities for the origin of this warming. One is global warming associated with increasing atmospheric greenhouse gases, such as carbon dioxide, whose effects are analogous to an atmospheric blanket. Simulations by climate-prediction models suggest that the warming will accelerate during the twenty-first century, with certain reductions in both lowland and mountain snow. Another contributor could be the Pacific Decadal Oscillation (PDO), a natural twenty- to thirty-year cycle in sea-surface temperatures and winds over the Pacific Ocean. The PDO was in its cool phase from roughly 1945 through the mid-1970s, when snowfall was greater and temperatures were lower, but switched in the late 1970s to the warm phase, which brings warmer temperatures and less snow. Chapter 12 of this book discusses the issue of global warming and its local impacts in greater depth.

Nearly all winter precipitation in the Northwest starts as snow aloft, irrespective of what reaches the surface, since the temperatures in the clouds are usually well below freezing. When temperatures remain below or near freezing beneath the clouds, snow can reach the surface, while if the temperature rises above 32 °F for at least 1,000 feet in the lower atmosphere, rain forms as the snow melts in the warmer air below. Many a Northwest air traveler can attest to this situation: landing at Seattle-Tacoma or Portland airports on a wet winter night, the movement of thousands of snow crystals is often apparent in the plane's lights a few minutes before touchdown, while on final approach the darting snowflakes are replaced by fast-falling raindrops.

PACIFIC NORTHWEST HISTORICAL SNOWSTORMS AND ICE STORMS

From historical records and newspaper accounts, it appears that the Pacific Northwest had a colder and snowier climate during the nineteenth and early twentieth centuries. As described in David Laskin's *Rains All the Time* and Walter Rue's *Weather of the Pacific Coast*, extended periods of below-freezing weather and heavy snow were noted between the 1830s and 1860s, with the Columbia River freezing for several weeks. Probably the coldest winter was 1861–62, when below-freezing weather began in mid-December and held on, with a few breaks, into mid-March. In his *Pioneer Days on Puget Sound*, Seattle founder Arthur Denny noted that 1861–62 was the "coldest winter we ever experienced on the Sound," with a record low of −2 °F. During that winter, up to 6 inches of ice covered Seattle's Lake Union and snow drifted to 6 feet during a particularly ferocious blizzard. The Columbia River froze, allowing horse-drawn sleigh and pedestrian passage between Oregon and Washington. Even heavier snow hit Seattle during January 6–7, 1880. Denny "made inquiry of the Indians and could get no account of anything like it before." He wrote that the snowfall measured "four feet and a half after it settled and would have measured much more as it fell."[2]

Seattle's greatest snowstorm since official record keeping began in 1890 occurred January 31– February 2, 1916 (figure 4.3). During that event, the city's daily snowfall record (21.5 inches) was established, with 29 inches on the ground by the time the storm was over. Outside of the city even greater snow depths were reported, with some wind-driven snowdrifts reaching 4–5 feet. Victoria, British Columbia, reported 4 feet of snow; Anacortes, Washington, received 2 feet; and 30 miles to the north of Seattle, the city of Everett lay under 3 feet of snow.

2   Arthur Denny, *Pioneer Days on Puget Sound* (1888; repr., Seattle: The Alice Harriman Co., 1908), available at the Washington State Library, www.secstate.wa.gov/history/publications.aspx.

4.3. Seattle after the record-breaking January–February 1916 snowstorm. Photo courtesy of the University of Washington Library, Special Collections (negative Todd97).

Numerous trees were downed, barns lost their roofs from the weight of the snow, and the dome of Seattle's Saint James cathedral collapsed. The effects of the February 1916 snowstorm extended south to northern Oregon, with Portland receiving 13.5 inches, Cascade Locks in the Columbia Gorge gaining 46.5 inches, and, east of the Cascades, Pendleton garnering a storm total of 28 inches.

The winter of 1949–50 stands as the coldest and snowiest of the twentieth century in both Oregon and Washington. Seattle-Tacoma Airport hit its all-time record low of 0 °F on January 31, while Portland Airport, in the path of cold easterly flow exiting the Columbia Gorge, set a record low of –3 °F on February 5. On January 13, 1950, 21.4 inches of snow fell on Seattle (the second greatest one-day total), with accompanying 25- to 40-mile-per-hour northerly winds that created blizzard conditions. Thirteen lives were claimed over the Puget Sound region and virtually all highways west of the Cascades were impassable. Seattle received 57.2 inches of snow that January, and for eighteen

4.4. The weight of heavy snow, saturated by intense rainfall, caused the collapse of roofs at an Edmonds, Washington, marina, destroying or damaging hundreds of boats. Photo courtesy of Mr. Chris Osterman of the Port of Edmonds.

days the mercury never climbed above freezing. In Oregon, a series of three storms brought record-breaking snows during January 1950, and a severe freezing-rain event on January 18 caused havoc on highways and felled power lines in and west of the Columbia Gorge. Total snowfall for January 1950 ranged from approximately 6 inches on the Oregon coast and 30 inches in the Willamette Valley to 40–50 inches east of the Oregon Cascades.

The winter of 1968–69 also brought extraordinary snow to the Northwest. Over much of the region, January 1969 was the coldest and snowiest month since 1950, with snow totals at many Northwest locations approaching or exceeding record amounts. Seattle received its all-time record winter accumulation of 67.5 inches, and Eugene in the Willamette Valley collected a record monthly snowfall of 47.1 inches, with 36.1 inches on the ground at one point—nearly triple the previous record of 11 inches. Along the coast, where the

average snowfall is generally less than 2 inches, January snow totals ranged from 2 to 3 feet, and Astoria, on the northwest Oregon coast, set a new monthly snowfall record of 25 inches.

Although the last two decades of the twentieth century brought a noticeable decrease in snow and cold at most Northwest locations, there were a few major snow events. Over Washington the most disruptive event occurred during the last week of December 1996, when two storms brought a combination of freezing rain and snow. The first storm, on December 26, dumped heavy snow north of Seattle, with 6–12 inches in many places. South of the city, this event produced the worst ice storm on record over the southern Puget Sound, resulting in the cancellation of hundreds of flights at Seattle-Tacoma Airport and making many streets impassable throughout western Washington. Due to heavy mountain snow, Snoqualmie and Stevens passes were closed for days, stranding motorists and severing US 2 and Interstate 90, the two major

4.5. One of the many landslides that occurred after heavy snow and rain struck the Northwest during the last week of December 1996. The combination of rapidly melting snow and heavy rain was equivalent to a hundred-year rainstorm, resulting in the failure of steep saturated slopes, such as this one on Bainbridge Island. Photo courtesy of the *Seattle Times.*

cross-state routes. Strong northeasterly winds exiting the Fraser River gap gusted up to 50 miles per hour, and wind chills approached −50 °F near Bellingham. Coupled with a 12-inch snowfall, these winds produced blizzard conditions and 10-foot snowdrifts. A day passed before the next storm hit late on December 28. Heavy snow fell over much of western Washington, with snow depths at many lowland locations reaching 18–24 inches before the snow turned to heavy rain between 6:00 and 9:00 AM on December 29. The combined weight of the snow and rain caused the collapse of hundreds of roofs, including dozens of boat shelters at the Edmonds marina (figure 4.4). Nearly four hundred boats sank as the heavy roofs pushed vessels underwater. The rapid melting of the large snowpack, in concert with heavy rain, resulted in 4–5 inches of liquid water hitting the ground over a period of about a day—the equivalent of the kind of rainstorm one expects to see only once every hundred years.

This massive and rapid influx of water produced widespread street flooding and the most extensive series of mudslides and slope failures of the past several decades (figure 4.5). Over eastern Washington, the second storm brought record-breaking snow that caused millions of dollars in damage, collapsing roofs of warehouses, fruit-storage facilities, schools, and other buildings. At Yakima the storm produced a record snow depth of 27 inches. Over all of Washington State, the December 1996 ice/snow/rain event killed twenty-four people, produced power failures that darkened over three hundred thousand homes, and resulted in damage conservatively estimated at five hundred million dollars. On the positive side, the December 1996 snowstorms were well forecast by the National Weather Service, a refreshing change from the poor snow predictions that had plagued the meteorological profession in the past.

WHY DO THE LOWLANDS OF WESTERN OREGON AND WASHINGTON RECEIVE SO LITTLE SNOW?

West of the Cascade Mountains, winter weather is usually mild and wet and occasionally cold and dry, but the combination of wet and cold is extraordinarily difficult to achieve. The lowlands of western Oregon and Washington typically receive only 5–15 inches of snow per year, and a winter season with little snow is not unusual; in contrast, cities closer to the equator, such as Washington, D.C. and Chicago, collect far greater annual snowfalls (23 and 38 inches, respectively). The explanation for the lack of snow over the Northwest lowlands is twofold: the relatively warm Pacific Ocean upstream of the region and the protective effects of the Cascade and Rocky mountains to the east.

The winds over the Northwest generally come from the west, and thus nearly all of the region's precipitation is associated with air that has passed over the Pacific Ocean, whose surface temperatures typically drop only into the upper 40s °F during midwinter over the immediate offshore waters. When dry, frigid air originating over Alaska or eastern Asia traverses the relatively warm ocean, it is moistened and warmed in the lowest few thousand feet. Thus, the temperature of the surface air reaching the Northwest coast during the winter is typically in the 40s °F, and lowland precipitation west of the Cascades usually comes in liquid form.

For the western lowlands of the Pacific Northwest to be cold enough for snow at the surface, the air generally has to come from the cold continental interior and thus must cross or pass through the substantial double barrier to the east, the mountains of the Cascades and Rockies (refer back to figure 2.2). Cold, dense air at low levels tends to flow around, not over, mountain ranges, and thus frigid air from northern Canada is typically deflected by the Rockies southeastward toward the Great Plains, not southwestward toward the Pacific Northwest. Occasionally, the cold air deepens sufficiently to pass over the Rockies and the Cascades, and in this case another protective mechanism is invoked: since air is compressed and warmed as it descends toward the higher pressure of lower elevations, its temperature is substantially moderated by the time it reaches the lowlands west of the Cascade crest. Thus, temperatures are substantially warmer in eastern Washington and Oregon than east of the Continental Divide, and are warmer yet over western Washington and Oregon after descending the Cascades. In addition to warming, cold air from the continental interior is usually quite dry and dries even more when it descends the western slopes of the Rockies and Cascades. This can be understood by noting that as air warms it can contain more water vapor. As a result, the relative humidity drops and clouds evaporate in the sinking air.[3]

---

3   As discussed in chapter 2, relative humidity is the ratio of the amount of water vapor the air actually holds divided by the maximum possible that it can hold at that temperature. At a relative humidity of 100 percent, the air becomes *saturated*, and clouds and precipitation can form. If the air warms, the amount of moisture the air can contain increases and thus the relative humidity drops. Since warming causes the relative humidity to drop, clouds tend to dissipate as air descends and warms along the lee slopes of a mountain barrier.

Fraser River Valley

Bellingham

4.6. View of the Cascades; Bellingham, Washington; and the eastern terminus of the Fraser River valley. Mount Baker is seen to the right. The gap in the Cascades surrounding the Fraser River is evident in the left-hand portion of the image. During cold-air outbreaks, frigid air pushing though the Fraser River valley can cause strong winds and blizzard conditions over and north of Bellingham. Graphic generated using Google Earth Pro, printed courtesy of Google Inc.

A limited amount of cold air *can* move westward through gaps in the protective mountain barriers. For western Washington, the Fraser River gap is the main low-level conduit across the Cascades and taps the frigid air within the interior of British Columbia (figures 2.2, 4.6), while for Portland and the northern Willamette Valley, the Columbia River gorge is the key cold-air passage. Rarely, the cold air of the continental interior is deep and frigid enough to push westward across the Rockies and Cascades, descend into western Oregon and Washington, and remain cold enough to support snow.

Significant lowland snow west of the Cascades is usually associated with cold continental air entering the region at low levels through such gaps or passes in the mountains, while moist air from the Pacific rises overhead, producing precipitation. The proper balance between cold low-level continental air and moist, but mild, Pacific air aloft is excruciatingly difficult to achieve: too much warm Pacific air and the snow turns to rain, while too much dry, cold air or sinking on the terrain can lessen or end the snow. In marginal situations, when air temperatures are in the mid-30s °F, the melting and evaporation of precipitation can determine whether snow or rain is observed near the surface. This is because both melting and evaporation can cool the air, sometimes sufficiently to turn rain into snow. Most of us have experienced such precipitation-related cooling immediately preceding or during the initiation of precipitation at the surface. Not surprisingly, the amount of melting and evaporation depends on the intensity of the precipitation, with heavier precipitation producing more cooling. This adds a major complication to the forecast problem, particularly since meteorologists have great difficulty forecasting precipitation amounts. It is easy to see why snow prediction is the most difficult forecasting problem of the Northwest lowlands: the conditions have to be just right and snow occurs so infrequently that even "gray-haired" meteorologists have only limited experience in its prediction.

SNOW SCENARIOS FOR THE PACIFIC NORTHWEST LOWLANDS

The majority of major snowstorms over the western Washington and Oregon lowlands evolve in a similar way, and thus much can be learned by reviewing a typical event. Consider the snowstorm of November 21–22, 1985, which brought 9.4 inches to Seattle-Tacoma Airport and similar amounts throughout the area. The first step in the evolution of this storm, as shown in figure 4.7 at 4:00 PM on November 20, begins with a cold, high-pressure area over northern British Columbia and the Yukon. Such Arctic cold-air masses generally form under high pressure and clear skies, as the snow-covered tundra or boreal forest radiates infrared radiation to space and cools progressively over a period of several days to a week. Clouds inhibit the loss of radiation to space and thus the greatest cooling occurs under cloud-free conditions that

accompany high-pressure regions. The upper-level chart in figure 4.7, which gives meteorologists an idea of atmospheric winds and pressure changes at approximately 18,000 feet above sea level, indicates an area of high pressure (known as a ridge) offshore and a low-pressure disturbance (known as a trough, indicated by a red dashed line in the figure) over southeast Alaska. Such upper-level troughs tend to cause clouds, precipitation, and the development of low-pressure systems in the lower atmosphere.

In the second step of the storm, at 4:00 AM on November 21, high pressure associated with the cold air as well as the upper-level trough had pushed southward into British Columbia. Low pressure had developed along the coasts of Washington and Vancouver Island, and the increasing difference in pressure between the interior of British Columbia and western Washington drew cold air westward through the Fraser River valley. The passage of cold air through the Fraser Valley is usually signaled by major changes at Bellingham, Washington: wind shifts rapidly to the northeast and strengthens to twenty miles per hour or more, temperatures decline, and relative humidity plummets. During such situations, Seattle TV stations, always looking for a way to hype the anticipated arrival of snow, often position their camera crews outside of Bellingham to show a wind-whipped correspondent with flags and trees blowing madly in the background.

In the storm's third step, at 4:00 PM on November 21, heavy snow is falling over western Washington. The disturbance aloft had strengthened and moved southward, contributing to the rapid development of a low-pressure system along the south-central Washington coast. As the low-

4PM 20 Nov   4AM 21 Nov   4PM 21 Nov

Upper level

Surface

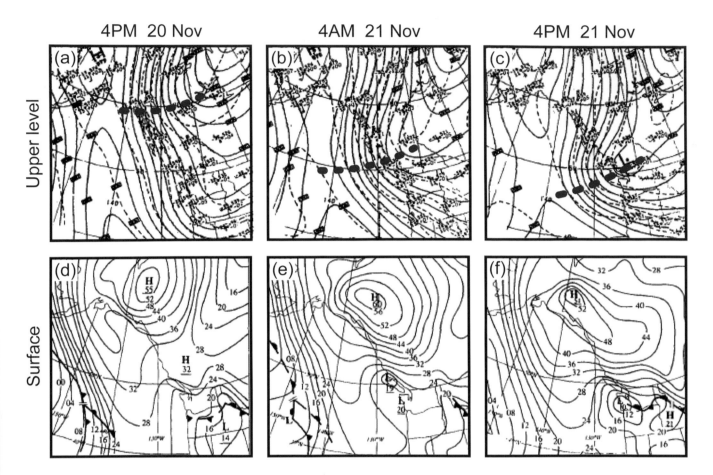

4.7. Upper-level (500-millibar) and surface National Weather Service weather maps during the western Washington snowstorm of November 21-22, 1985. The lines on the upper-level map are parallel to the winds at approximately 18,000 feet, with the red dashed line indicating the location of a weather disturbance aloft (known as a trough). The surface chart indicates sea-level pressure, with the lines being isobars (lines of constant pressure). "H" and "L" indicate areas of relatively high and low pressure, respectively.

pressure center deepens, the difference in pressure across the Fraser River gap in the Cascades increases, greatly strengthening the influx of cold air. Winds at Bellingham can gust to fifty or sixty miles per hour from the northeast in such situations, with most of this strong current flowing across the San Juan Islands and out the Strait of Juan de Fuca, and a smaller portion heading southward toward Puget Sound. A trip to the top of Mount Constitution on Orcas Island reveals numerous trees leaning to the southwest or toppled in that direction due to the strong northeasterly Fraser outflows of the past. Moist air from off the ocean circulates around the coastal low and then rises as it moves over the low-level cold air that had moved southward out of British Columbia (figure 4.8a). The boundary between the cold continental air from the interior and the relatively warm air from off the ocean is known as the *Arctic Front*. One can visualize the Arctic Front as a cold air wedge that pushes warm air aloft. As described

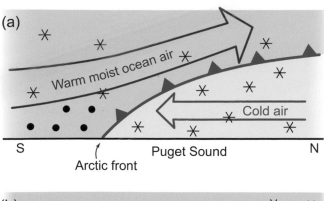

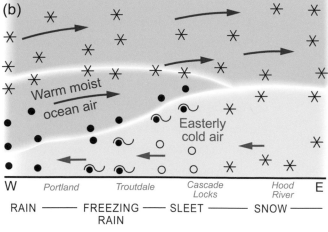

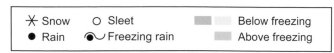

4.8. (a) Schematic of the weather situation associated with many snowstorms over western Washington. Cold, dense air, much of it coming through the Fraser River gap of British Columbia, pushes south, while moist, relatively mild air from off the Pacific ascends over the cold air. The boundary between the two air masses is often called the Arctic Front. (b) Schematic of the complex weather situation associated with frozen precipitation in and near the Columbia Gorge. As the depth of the cold air decreases toward the west, snow transitions first to sleet, then freezing rain, and finally to rain where the depth of the subfreezing air is shallow. Illustration by Beth Tully/Tully Graphics.

in chapter 2, rising air enhances precipitation, and thus heavier precipitation is usually noted at or north of the Arctic Front. In the warmer air south of the Arctic Front, snow from aloft melts into rain as it approaches the surface, while north of the Arctic Front snow can reach the surface.

Although perhaps 80 percent of western Washington snowstorms occur in the manner described above, there are other possibilities. Sometimes cold air and an area of high pressure are already in place, with the region enjoying an unusual period of cold, sunny weather.

Subsequently, the atmospheric circulation shifts so that the cold high-pressure air is invaded on its southern and western flanks by a moist, westerly air current associated with a Pacific low-pressure center. As the Pacific low-pressure system moves eastward toward the region, more low-level cold air is drawn into western Washington as the pressure difference across the region increases (high pressure inland, lowering pressure over western Washington). At the same time, the approaching disturbance pushes relatively warm, moist air overhead. The result can be a very heavy, wet snowfall.

In most cases such snow soon turns to rain as the atmosphere warms, leaving a slushy mess and often flooding. The December 28–29, 1996, storm was such a snow-turning-to-rain event, with water-laden snow collapsing roofs and the combination of rapidly melting snow and heavy rain resulting in slope failures throughout the region.

LOCALIZED SNOWSTORMS AND ICE STORMS

Some snowstorms over the Northwest lowlands are surprisingly localized, with heavy snow at one location and not a flake a few miles away. These local events are usually associated with regional terrain features. For example, cold, northeasterly flow pushing out of the Fraser River valley some-times crosses the San Juan Islands, picks up mois-ture over the Strait of Georgia and the eastern Strait of Juan de Fuca, and then rises over the northeast slopes of the Olympics, producing snow between Sequim and Port Angeles—quite a shock to those living in a region known for its rain shadow (figure 4.9)! Localized snow can also occur on the south-east slopes of the Olympics and nearby Kitsap County during periods of strong southeasterly flow, even when temperatures are borderline for low-level snow over the region. As the air moves up the southeast slopes of the Olympics, it is forced to rise and cool, producing heavy precipitation. If temper-atures start out a few degrees too warm for snow near the surface, the cooling of rising air and the melting of snow falling from aloft progressively causes the temperature, and thus the freezing and snow levels, to drop—until snow, often heavy, reaches the surface. Since the influence of the Olympics can extend tens of miles upstream of

the barrier, snow can extend well into Kitsap County, affecting the region around the Hood Canal and towns such as Bremerton and Belfair.

Another local snow producer is the Puget Sound Convergence Zone, which occurs when winds from the west are deflected around the Olympics and converge on the eastern side of that barrier, pro-ducing a narrow band of clouds and precipitation (see chapter 7 for more details on the convergence zone). During winter periods when the tempera-ture is marginal for snow, the heavy precipitation in the convergence zone causes intense cooling in the lower atmosphere as the snow crystals falling from aloft are melted in the warmer air below. This cooling can drive the snow level to the surface in the heavy convergence zone precipitation. North or south of the heavy convergence-zone showers, where the precipitation is lighter and thus cooling by melting is less, rain is observed. Since the con-vergence zone is most prevalent between Seattle and Everett, this area is the most vulnerable to sur-prise convergence zone snow events.

One of the most dramatic convergence zone snowstorms occurred on December 18, 1990, an event that was poorly forecast. A cool air mass had begun to spread over the region aloft, but at low levels the Arctic Front and coldest air were still well to the north. A convergence zone formed over Puget Sound during the morning and proceeded to intensify, with rapidly developing cumulus and cumulonimbus (thunderstorm) clouds producing heavy snow showers, lightning, and thunder over central Puget Sound. While parts of Seattle and Bellevue received a foot or more of unpredicted snow, areas to the north of Everett or south of Seattle received virtually nothing (figure 4.10). Traffic ground to a standstill, with buses and cars

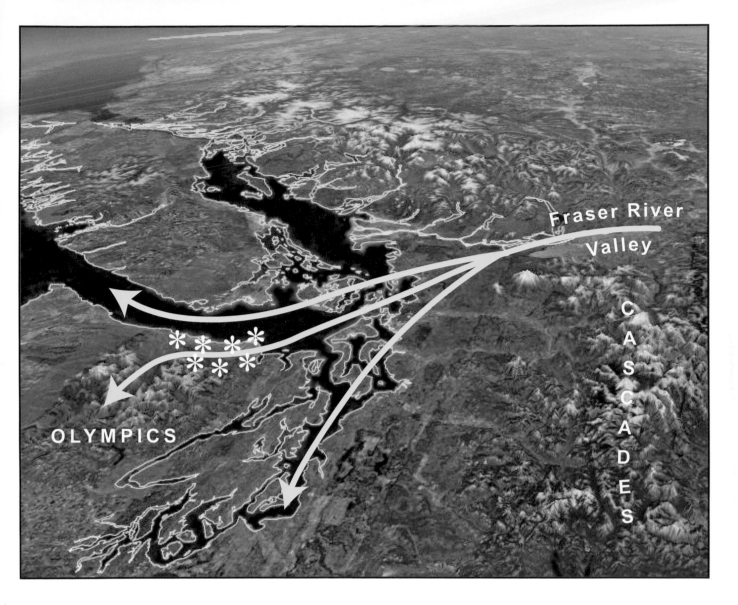

Fraser River
Valley

C
A
S
C
A
D
E
S

OLYMPICS

4.9. Cold air moving southwestward out of the Fraser River valley is often forced to rise by the Olympic Mountains, producing snow on the northeast portion of the Olympics that can spread over the normally rain-shadowed town of Sequim. Graphic generated using Google Earth Pro, printed courtesy of Google Inc.

abandoned throughout Seattle and the eastern suburbs. Later in the day, the Arctic Front pushed across Puget Sound, bringing much colder temperatures and strong winds that downed power lines and created icy conditions. The author, having failed to forecast the snow and temporarily trapped at the University of Washington, retreated to a local restaurant for dinner with other faculty and staff. As the Arctic Front slammed into the city, the windows of the restaurant suddenly fogged over as the windows rapidly chilled, causing the water vapor in the restaurant to condense on the glass.

Lower-elevation snow in western Oregon is also relatively rare. The situations producing snow in western Oregon are generally similar to those of western Washington, except that the most important low-level source of cold air is not the Fraser

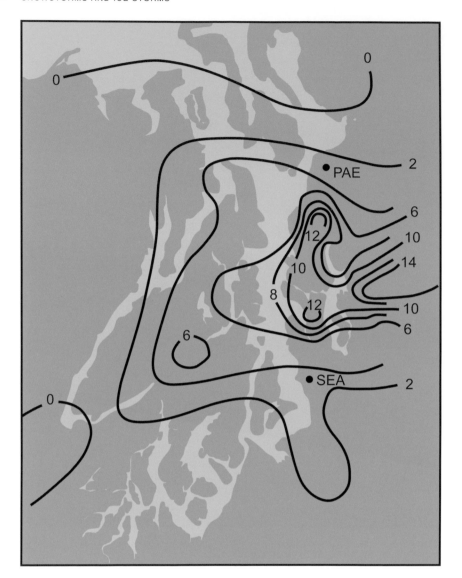

4.10. Snow depths after the surprise convergence zone snowstorm of December 18, 1990. Amounts shown in inches. SEA is Seattle-Tacoma Airport; PAE is Paine Field in Everett.

River valley, but the Columbia River gorge. The gorge affords a near sea-level passage through the Cascade Mountains for cold air from the interior of eastern Washington. This cold air often originates in British Columbia and reaches the "bowl" of the Columbia Basin through the Okanogan and adjacent valleys. The typical snow scenario for Portland and the nearby northern Willamette Valley begins with cold air and high pressure to the east of the Cascades and lower pressure along

the coast; this difference in pressure forces cold air westward through the Columbia Gorge, producing easterly winds at gorge locations such as Cascade Locks and Troutdale that can extend westward to Portland and beyond. As a Pacific low-pressure system approaches the region, the cross-Cascade pressure difference is enhanced, increasing the cold easterly flow though the Gorge, while at the same time moist, Pacific flow moves in overhead. With snow generated aloft falling into cold air at low levels, snow can reach the surface over the Gorge, Portland, and the northern Willamette Valley (see figure 4.8b). If the cold air east of the Cascades becomes sufficiently deep, it can push across the Cascade passes and down the many river drainages along the western slopes, contributing to more widespread lowland snow.

Sometimes the approaching Pacific disturbance brings in moist air aloft that is sufficiently warm so that snow forming above falls into an above-freezing layer, causing the snow to melt into rain. If the rain then falls into subfreezing easterly gorge flow near the surface, it can be cooled below freezing. Interestingly, even if the temperature of the rain falls below 32 °F, it does not necessarily freeze immediately, but can remain liquid as *supercooled* water. When such supercooled water hits a cold surface or solid object, it can freeze on contact, producing a glaze of ice that coats and downs power lines or makes travel treacherous. Portland and the eastern Columbia Gorge often experience such freezing-rain events, which are sometimes called silver thaws. To produce freezing rain, the depth of the subfreezing cold air has to be just right: too deep and the rain freezes solid into sleet before it hits the ground, too shallow and the rain is not cooled to 32 °F or below and falls as ordinary rain. Often the depth of the cold air varies across the Gorge, thinning as it fans out west of the gap. In such situations the central Gorge gets snow, Troutdale and environs receive sleet, Portland experiences freezing rain, and rain falls farther to the west.

One of the most extraordinary and long-lived Portland snow and ice storms on record struck during January 6–9, 2004. Strong high pressure built up east of the Cascade Mountains on January 5,

forcing frigid air westward through the Columbia River Gorge into northwest Oregon. A Pacific low-pressure system spread moist air above this cold easterly flow, resulting in widespread snow, sleet, freezing rain, and blizzard conditions in the Gorge. Snowfall ranged from 2 to 3 inches along the northern Oregon coast and from 2 to 8 inches in the Willamette Valley to 27 inches in the Cascades. Accumulations of up to 2 inches of sleet and freezing rain followed the snow over the Portland metropolitan area, as the air aloft began to warm and caused rain to fall into the subfreezing air near the surface (figure 4.11). A thick veneer of ice closed the Portland Airport for a record three days, resulting in the cancellation of over 1,300 flights and stranding ninety thousand passengers. Portland's light-rail system was shut down and most businesses and schools were closed for several days. Blizzard conditions halted east-west travel in the Columbia Gorge, stranding hundreds of trucks: Interstate 84 closed between Troutdale and Hood River, Oregon, and State Route 14 closed between Washougal and White Salmon, Washington. The weight from snow and ice caused widespread downing of trees and power lines, leaving forty-six thousand customers without power, and collapsed roofs at Portland's Gunderson Steel and Fred Meyer department store.

Small differences in elevation or proximity to water can have a large effect on snowfall amounts over the western lowlands, where surface temperatures during winter are often borderline for snow. Since water bodies, such as the Pacific Ocean, Puget Sound, the Strait of Juan de Fuca, and Lake Washington, only cool into the 40s °F during the winter, the air near the surface is usually above freezing, and locations near or immediately downwind of the water are often too warm for snow. The

(a)

(b)

4.11. An ice storm that struck Portland, Oregon, in January 2004 coated the entire metropolitan area with several inches of ice and resulted in a record three-day closure of Portland International Airport. Photos courtesy of (a) Tyree Wilde, warning coordination meteorologist, National Oceanic and Atmospheric Administration/National Weather Service and (b) Dr. Justin Sharp.

4.12. Seattle's Capitol Hill on March 25, 2002, looking south from the University of Washington. A very sharp transition to snow occurred at around 150 feet above sea level.

ington or Puget Sound, but wet snow a mile or so away on top of View Ridge, Capitol Hill, or Queen Anne Hill (figure 4.12).

HEAVY SNOW IN THE CASCADES

Even though the Pacific Northwest lowlands are usually snow-free, the Cascades often receive heavy accumulations, with some locations buried under 50–100 feet of snow over a winter season. The substantial Cascade snowfall is not only important for winter recreation, but also serves as a crucial source of water for human consumption, irrigation, hydroelectric generation, and fisheries. Why is it that so much snow can fall in the mountains, while the lowlands are rarely visited by snowflakes?

tops of the higher urban hills of Seattle or Portland, which typically reach 300–500 feet, can be 1–3 °F cooler than at sea level; in marginal situations, this small temperature drop can make the difference between rain and snow. In Seattle, it is not unusual for there to be rain at the elevation of Lake Wash-

The air approaching the Northwest, often originating over Siberia or Alaska, is both moistened and warmed in the lower few thousand feet by its long journey over the relatively mild Pacific Ocean, with wintertime air temperatures near sea level typically in the 40s °F. As described earlier in this chapter, the western lowlands rarely get snow with such mild temperatures. Temperature decreases with height in this Pacific air, generally by approximately 3 to 5 °F per 1,000 feet. Thus, on a typical winter's day with a maximum temperature of 44 °F in Seattle, the temperature at 4,000 feet would be expected to reach approximately 28 °F, with precipitation falling in the form of snow. In fact, since snow does not melt instantaneously when it encounters above-freezing temperatures, wet snow can occur at temperatures a few degrees above freezing. Under typical conditions, the *snow level* is about 1,000 feet below the *freezing level*, the height in the atmosphere where the air temperature equals 32 °F. Thus, with surface temperatures in the mid-40s over the Oregon or Washington lowlands, snow is generally found above 3,000–4,000 feet along the western slopes of the Cascades.

The heaviest snow in the Cascades usually does not occur while fronts and low-pressure areas are crossing our region, but rather after they pass to the east, as cool unstable air floods the area and winds develop a westerly component. As shown in figure 4.13a, when a low-pressure system approaches the Northwest coast the winds are generally from the south. Such southerly flow brings warm air that can raise the freezing level, sometimes as high as 7,000–8,000 feet. Since the Cascades are oriented roughly north-south, southerly winds are not lifted much by the Cascades, thus lessening the amount of precipitation that is

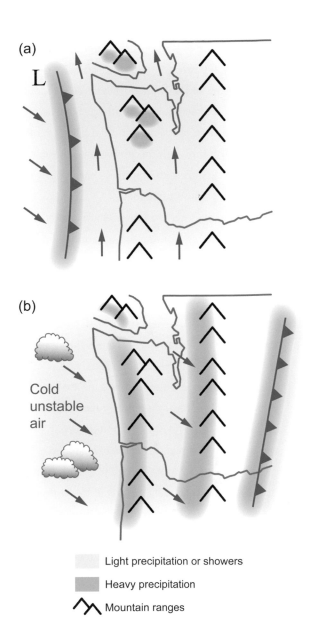

4.13. (a) Before Pacific fronts make landfall, the winds at and below mountain-crest level are often southerly, with relatively little upslope motion, high freezing levels, and light mountain precipitation. (b) In contrast, after frontal passage, the conditions are ideal for heavy mountain snow, with a cold, unstable air mass and a wind direction nearly perpendicular to the mountain slopes, resulting in large upward motion. Illustrations by Beth Tully/Tully Graphics.

released. Only on south-facing slopes, such as the southern Olympic Mountains, do southerly winds result in heavy precipitation. In short, with warm temperatures and only a modest amount of mountain-induced precipitation, southerly flow, as commonly occurs prior to Pacific frontal passage, often brings only minor snow accumulation or even some melting of preexisting snow, particularly on lower slopes below 4,000 feet.

After the passage of a low-pressure center and its associated front(s), the situation changes dramatically (figure 4.13b). Falling temperatures behind the system result in plummeting freezing and snow levels—particularly above the layer influenced by the ocean surface. With colder air above and Pacific-warmed air below, the change in temperature with height increases. As a result, the atmosphere becomes unstable, producing convective clouds such as cumulus or cumulonimbus and showery precipitation. As noted in chapter 2, a similar phenomenon occurs in a saucepan of hot cereal, where a large change of temperature in the vertical, caused by heating at the bottom of the pan, produces rising and descending convective currents in the cereal. As the weather system moves across the region, the winds tend to shift from southerly or southeasterly to a more westerly direction, pushing air directly up the slopes and thus producing strong upward motion over Cascade slopes. Such upward motion enhances the showery convective clouds, producing copious precipitation in the form of snow if the air is sufficiently cold. For all of these reasons, it is after low pressure and frontal passage that the greatest snowfall occurs on the western slopes and crests of the Cascades, occasionally with several feet in a day.

The major passes in the Cascades, representing significant gaps in the mountain range, enjoy far more snow than expected for their lower, warm elevations. Without such gaps, which offer a conduit for colder air from the east, skiing would not be viable at moderate elevations (3,000–4,000 feet) in the Cascades. For example, winter recreation at the Summit ski area at Snoqualmie Pass, with its base near 3,000 feet in the central Washington Cascades, is possible only because cold, easterly flow from eastern Washington often protects the pass from the warm air that precedes Pacific storms and fronts. Skiing and snowboarding are not feasible at the same elevations on the western side of the Cascades because temperatures are too warm. In fact, an attempt was made to run a ski area at Mount Pilchuck on the western slopes of the Cascades between 2,500 and 4,300 feet.[4] Although this facility was open during most years from 1957 to 1980, a cool period of relatively plentiful snow over the region, it was forced to close in 1980 due to poor snow conditions. Such a fate may be in store for even pass locations like Snoqualmie as global warming causes the freezing and snow levels to rise over the next fifty years.

The easterly flow that supports winter recreation in Snoqualmie and Stevens passes is driven by the difference in pressure between cold, dense air east of the Cascades—and associated high pressure—and low pressure along and off the coast. As shown in figure 4.14a, this easterly flow is often quite shallow, with a sharp transition to warmer southerly or westerly air aloft. Occasionally, the

---

4   Mount Pilchuck is roughly 20 miles northeast of Everett, Washington.

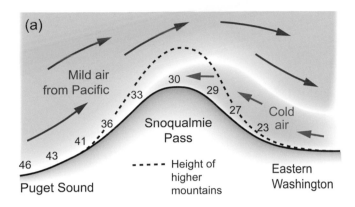

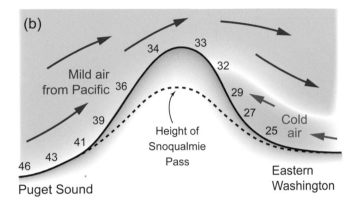

4.14. (a) Cross-section of Snoqualmie Pass with cold air from eastern Washington moving westward through the pass, resulting in temperatures cold enough for snow. (b) In contrast, in other areas of the Cascades where a low-elevation pass does not exist, the cold air is generally not deep enough to cross the barrier. Thus, locations on the western side of the Cascades at similar elevations to Snoqualmie Pass are too warm to build a significant snowpack. Illustrations by Beth Tully/Tully Graphics.

lower slopes of the ski areas are in the teens or 20s °F, while the upper slopes are 10–20 °F warmer. If the air aloft warms up sufficiently, rain can fall into subfreezing cold air in the passes, producing treacherous freezing rain that often contributes to accidents. In contrast, where a pass does not

exist (figure 4.14b), cold air is usually trapped on the eastern side of the mountains and the lower western slopes of the Cascades experience only the relatively warm air from off the Pacific Ocean, producing poor conditions for snow accumulation.

In general, snowfall tends to increase with height in the mountains, but there is an elevation above which snowfall actually starts to *decrease*. For example, a number of climbers have noted that more snow falls at Mount Rainier's Paradise Ranger Station (approximately 5,500 feet above sea level) than at Camp Muir (10,000 feet). There are a number of possible explanations for a decline in snowfall at the highest elevations. One reason is that the amount of water vapor in the air decreases with height (colder air holds less water vapor), so that air moving up an upper mountain slope produces less precipitation. Another factor is that the amount of terrain available for blocking the flow (and thus making it rise) declines at the highest elevations—in other words, it is easier for air to move around than over the upper portions of a mountain barrier.

SNOWSTORMS AND ICE STORMS EAST OF THE CASCADES

Isolated from the temperate influence of the Pacific by the Cascade Mountains and at a higher elevation than the western lowlands, eastern Washington and Oregon are substantially colder than their western counterparts and experience more snow during the winter. As described in chapter 2, snow east of the Cascades is greatest over the high Oregon Plateau and the mountains of northern Washington and least in the low-elevation Columbia Basin.

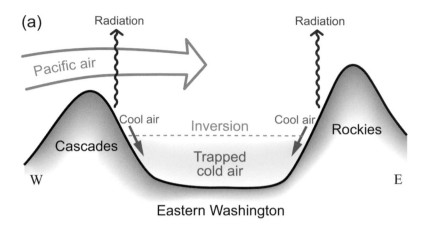

(a)

Radiation

Radiation

Pacific air

Cool air

Inversion

Cool air

Rockies

Cascades

Trapped
cold air

W

E

Eastern Washington

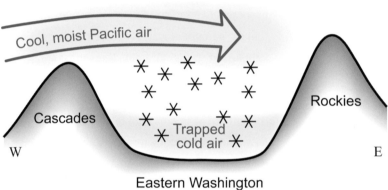

(b) Snow

Cool, moist Pacific air

Rockies

Cascades

Trapped
cold air

W

E

Eastern Washington

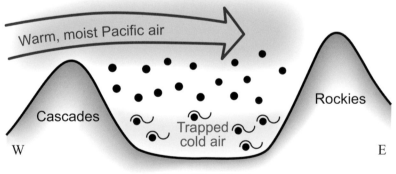

(c) Freezing rain; ice storms

Warm, moist Pacific air

Rockies

Cascades

Trapped
cold air

W

E

Eastern Washington

✳ Snow   ● Rain   ⊙∿ Freezing rain

4.15. (a) Configurations of cold air within the basin of eastern Washington. Cold air is often trapped at lower elevations, with warmer Pacific air overhead. (b) If the air aloft is relatively cold, then snow can occur over eastern Washington. (c) If some warmer air moves in aloft over eastern Washington, freezing rain or sleet can occur. Illustrations by Beth Tully/Tully Graphics.

Eastern Washington is essentially a large topographic bowl or basin, with the arid Columbia Basin near the Tri-Cities (Pasco, Richland, and Kennewick) being only a few hundred feet above sea level. This bowl is surrounded by the Cascades, the Okanogan Highlands, the Rockies, the Blue Mountains, and the higher elevations of the Oregon Plateau, with only the narrow Columbia River gorge acting as a drain. During the winter, cold air—either produced by radiational cooling[5] on the basin's slopes or entering through valleys and gaps in the surrounding mountains—is often trapped in the bowl for extended periods of time (figure 4.15a). Usually there is warmer air above, and a very stable layer, called an *inversion*,[6] that separates the cold low-level air from the warmer air aloft.

When a Pacific weather system moves across the region during the winter, the type of precipitation in eastern Washington varies with the temperature and depth of the low-level cold air in the basin and the temperature of the air aloft. When the air aloft and near the surface is relatively cold, then snow is observed at the ground (figure 4.15b), but when a layer of above-freezing air is found aloft, rain can fall into a subfreezing layer of air near the surface, resulting in freezing rain or ice pellets (sleet) (figure 4.15c). If the low-level cold air is absent or very thin, and warm air moves in aloft, rain can reach the surface.

Frequently, there is a mix of snow, freezing rain, and rain east of the Cascades that changes with elevation and the depth of the cold air. For example, on a number of occasions there is snow over the western side of the Columbia Basin at locations such as Wenatchee and Yakima, while freezing rain or rain is occurring over the eastern side at Spokane, Pullman, or Walla Walla. This contrast is often associated with *cold air damming*, whereby the cold air is deepest along the eastern slopes of the Cascades and thins dramatically toward the east (figure 4.16). Cold-air damming usually occurs when there is cold Arctic air and high pressure to the north. With higher pressure to the north and lower pressure to the south, the cold air tends to push southward into eastern Washington through valleys in the Okanogan Highlands. The cold air then banks up along the eastern slopes of the Cascades, resulting in the coldest, densest air being most thick on the western side of the basin.[7] Such a deeper layer of cold dense air produces a tongue of higher pressure over the eastern side of the barrier. Deeper cold air on the western side of the basin maintains snow in that area and also helps supply cold air to the lower passes in the Cascades, thus supporting skiing at Snoqualmie and Stevens passes.

---

5   The earth, like all objects, emits radiation that can escape to space, thus cooling the surface. Radiational cooling is most effective under clear or nearly cloud-free conditions.

6   In an inversion, temperature increases with height, the opposite of the normal cooling with height.

7   The cold air is pushed westward up the eastern slopes of the Cascades by the Coriolis force, a force that occurs because we are on a rotating planet. In the northern hemisphere the Coriolis force tends to push objects (or air) to the right of their direction of motion. So if winds are blowing from the north, the Coriolis force will provide a push to the right (westward) or up the eastern slopes of the Cascades, thus thickening the cold air layer there. The Coriolis force is discussed further in chapter 5.

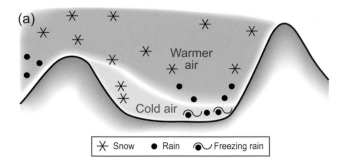

(a)

* Snow   • Rain   ◖◡ Freezing rain

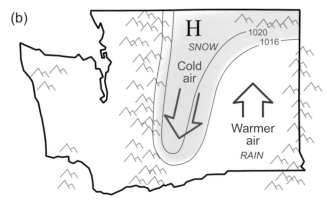

(b)

H

1020
1016

SNOW

Cold air

Warmer air
RAIN

Cold Air Damming

4.16. With cold air and high pressure to the north, cold air is usually deeper on the western side of eastern Washington. This phenomenon is termed *cold air damming* and is reflected in a tongue of high pressure over the eastern slopes of the Cascades. Cold air damming helps supply cold air to the Cascade passes and increases the likelihood of snow over the western side of the eastern Washington basin. Illustrations by Beth Tully/Tully Graphics.

One of the most memorable eastern Washington snow and freezing-rain events occurred during November 18–19, 1996. During the previous days, Arctic cold air and high pressure first invaded British Columbia and then infiltrated into eastern Washington. A low-pressure system moving eastward along the Washington-Oregon border then pushed warm, moist air northward over the region. On the western side of the basin, cold-air damming and northerly flow maintained a sufficient depth of cold air so that record amounts of snow fell on the eastern Cascade slopes and adjacent areas, with 14 inches at Yakima, 18–22 inches at Ellensburg, 13.5 inches at Wenatchee, and 15–22 inches at Goldendale. Fifteen thousand homes lost power in Yakima, several roofs collapsed there, and Interstate 82 was littered with over two-dozen jackknifed trucks.

To the east, the cold air layer thinned, resulting in freezing rain and rain. Freezing rain quickly collected on trees and power lines and eventually became heavy enough to break tree limbs (figure 4.17), with the ice up to 3 inches thick. Many of these tree limbs fell on power lines, causing approximately a hundred thousand Spokane County residents to be without electricity; some customers were without power for as long as nine weeks. Ten deaths and over twenty-two million dollars of damage were attributed to the ice storm. Washington governor Mike Lowry declared a state of emergency in Spokane County, followed by President Clinton naming the region a federal disaster area.

A little over a month later, during December 28–30, 1996, eastern Washington was again struck by a record-breaking winter weather event. Yakima, with 27 inches of snow on the ground, broke all-time daily and annual records for snowfall and snow depth. Yakima County officials estimated that such a snow accumulation would only be expected once in two hundred years. The weight of the snow caused the failures of hundreds of warehouse and building roofs. Major roadways such as Interstate 82 from Yakima to Ellensburg were closed, and in Chelan an overnight snowfall of 14 inches collapsed the roofs of schools and fruit-packaging plants. Damage to buildings was esti-

4.17. The ice storm of November 18-19, 1996, resulted in up to 3 inches of ice on trees and structures over Spokane and neighboring communities. Photo courtesy of Dan Hagerman of Spokane.

mated to exceed thirty million dollars. The heavy snow extended up into the Okanogan Valley, where a blizzard dumped several feet of snow that closed all state routes for two days.

PACIFIC NORTHWEST SNOW AND EL NIÑO AND LA NIÑA

Although the correlation is not perfect, there is a significant relationship between surface temperatures of the tropical Pacific Ocean and the amount of snow that falls in the Pacific Northwest. The sea-surface temperatures in the central and eastern tropical Pacific tend to swing between warmer than normal temperatures (known as El Niño) and cooler than normal temperatures (La Niña), with the intermediate years known as neutral years. Typically, this cycle takes three to seven years and is known in the weather business as the El Niño Southern Oscillation, or ENSO (chapter 11 elaborates on El Niño and La Niña).

A major advance in understanding occurred in the 1980s when it was realized that this variation in tropical Pacific sea-surface temperatures had a noticeable impact on the weather outside of the tropics, including the Pacific Northwest. During El Niño years, Northwest winters tend to be warmer than normal and precipitation near or slightly below normal, with the result being less snow over both the lowlands and the mountains. In contrast, during La Niña years Northwest temperatures tend to be cooler than normal, while precipitation is generally above normal. Not surprisingly, Northwest snowfall tends to be greater during La Niña years at all elevations. For example, from 1970 through 2000 at Seattle-Tacoma Airport, El Niño years typically brought 5 inches of annual snow, compared to an average of 15 inches during La Niña years. Paradoxically, although neutral years receive near-normal snowfall on average, record annual snowfalls have generally occurred in such years. Since the tropical ocean temperatures change relatively slowly, during the late summer and early fall one can often get some insight into the character of the winter's snowfall if one knows the current status of the tropical Pacific. ENSO-like variations also appear to occur on decadal time scales, and the corresponding Pacific Decadal Oscillation (PDO) also correlates with Northwest weather. As described in chapter 12, the cool phase of the PDO, peaking in the 1950s and 1960s, was associated with greater snowfall in the Northwest.

ROADWAY ICE—THE GREATEST WINTER WEATHER THREAT IN THE PACIFIC NORTHWEST

Ice on roadways is probably the most serious meteorological hazard in the Northwest and causes hundreds of injuries and several deaths each year. For example, Washington State Department of Transportation statistics for the region including the Cascade passes and the east side of the state indicate that roughly 25 percent of the total accidents are related to ice on the road. Sometimes called black ice when not clearly visible at night, roadway ice is not black at all, but is made up of frozen water that reflects light or sparkles when illuminated at the right angle (figure 4.18). As described below, roadway icing occurs under conditions that are generally well understood and often predictable.

Frost

Frost tends to occur on cold, clear or nearly clear nights when wind speeds are less than approxi-

4.18. Roadway icing, sometimes called black ice, is associated with many traffic accidents over the Pacific Northwest. Snow falling during rush hour on November 27, 2006, melted and then froze into slippery ice over much of the central Puget Sound region. Photo courtesy of the Washington State Department of Transportation.

mately 7 miles per hour. Why are clear skies and light winds important? All objects give off or emit infrared radiation, with warmer objects emitting more radiation. We all have some experience with infrared radiation; for example, when you sit across a room from a fire you can feel the infrared radiation it emits. Some objects emit infrared radiation better than others. For example, the earth's surface and clouds are far more efficient in emitting (and absorbing) radiation than are the gases in the atmosphere. On a clear night, the earth's surface emits infrared radiation, and with no clouds to stop it, much of this radiation is lost to space. As a result, the surface and adjacent air cool quickly. Thus, on cold, clear nights temperature often *increases* with height in the lower atmosphere, the opposite of the normal situation, and thus is known as an inversion (figure 4.19). On overcast nights, clouds absorb infrared radiation leaving the surface and send some back to the ground. Thus, clouds act like meteorological blankets that lessen the loss of infrared radiation from the surface and thus reduce surface cooling. If temperature increases with height, such as during inversion conditions, the infrared radiation from clouds can even *warm* the surface. Strong winds also tend to work against surface cooling, since windy conditions "stir up" the atmosphere and mix down some of the warmer air aloft.

Water vapor is an invisible gas and is seen only when it condenses into water droplets or ice crystals. As noted earlier, the amount of water vapor air can hold varies with temperature, with warmer air able to contain more water vapor. If we cool air down sufficiently (to what is known as the *dew point temperature*), it can no longer hold the moisture it started with, and thus some of the water vapor condenses into water droplets or ice crystals.

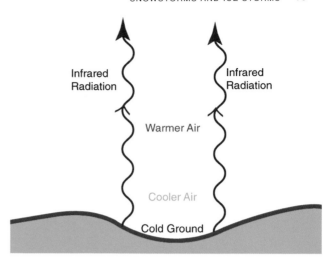

4.19. The surface emits infrared radiation to space more effectively than the air aloft, resulting in the development of an inversion in which temperature increases with height. Clouds interfere with the surface cooling because they intercept this radiation and emit some back to the surface.

During the day when the sun is out and warms the ground, the air temperature near the surface is usually above the dew-point temperature, and water in the atmosphere remains in the form of invisible vapor. However, as the sun sets on cold, clear nights, the surface temperature plummets as the earth continues to emit infrared radiation into space, allowing the air near the surface to cool to the dew-point temperature. If the roadway temperature and dew-point temperature are above freezing, liquid water forms on the surface (producing dew), but if the temperature is below freezing, frost forms instead. Frost is often more prevalent in valleys and low areas, where cold air tends to drain and pool (more details on this below, in the discussion of local terrain effects on roadway icing).

Frost generally accumulates slowly and generally only forms a thin layer of ice. For that reason, frost is generally less of a threat than fog-related and other forms of icing. But keep in mind, frost-related roadway icing has caused plenty of accidents!

Fog and Icing

Although the icing threat from frost is lessened by its slow accumulation and relatively minimal thickness, this is not true of fog-related roadway icing. Fog often forms on cold, clear nights as the air temperature drops to the dew-point temperature and there is enough mixing to cool down a layer of air (tens to hundreds of feet thick) to the dew point. Fog contains large amounts of liquid water, and if a fog bank passes over a roadway that has cooled below freezing, icing can be rapid and severe, with a thick coat of ice being deposited in minutes.

A number of serious icing accidents have occurred in the Northwest as a result of fog-related icing. A typical scenario starts with a clear, cold night in which the surface rapidly cools. A light frost might form on the roadway, but nearby fog begins to form over a moist surface, such as a river or swampy area. The fog drifts over the road, and as it passes over the cold surface a coating of ice is deposited. So you should be very careful if driving on a cold night if fog is nearby.

Other Causes of Northwest Roadway Icing

Snow can cause roadway icing, particularly after a period of above-freezing temperatures. The snow on the road starts to melt, but as temperatures drop, perhaps following passage of an Arctic Front or during the early morning hours on a clear night, the melting snow freezes, creating a thick layer of slippery ice. The former situation occurred on November 27, 2006, when an intense snow band associated with an Arctic Front dropped a few inches of snow over central Puget Sound between 4:00 and 5:00 PM during evening rush hour. Initially melting, the snow/slush mixture quickly froze as frigid air behind the Arctic Front moved in, making travel treacherous, particularly on hills and inclines. Traffic was brought to a standstill, as cars and buses became stuck on major roadways and offramps (figure 4.18). The commute home took hours for most people, and some abandoned their cars and walked home. Melting of snow pushed to the sides of roads can also cause roadway icing, as the resulting water runs over the road and freezes at night.

Freezing rain can also make roads icy, particularly over the Columbia Gorge area, the Cascade Mountain passes, near Bellingham, and over much of eastern Washington. As noted earlier in this chapter, freezing rain occurs when there is a layer of below-freezing air near the surface with above-freezing air aloft. Rain from aloft falls into the cold layer, cools below freezing (staying liquid as supercooled water), and then freezes immediately upon contact with the surface or some object.

The Effects of Local Terrain and Land Use on Roadway Icing

Roadway icing in the Northwest can be very localized, with the transition from safe to dangerous driving often occurring over small distances. Of particular importance is the tendency for cold air and ice to be found in low spots. Cold air is denser and heavier than warm air, and thus the coldest air tends to drain into low areas and valley bottoms (figure 4.20). In such situations, changes in temperature with elevation can be quite large, with the surface air in even shallow valleys (100–200 feet deep) being 2–5 °F cooler than air above higher ground a few hundred feet away. Clearly, the potential for roadway icing is greatest in valleys and low spots, which can be ice covered even when slightly higher elevations are ice-free.

At night, surface temperatures are usually warmer and roadway icing is less prevalent in cities than in surrounding rural areas, since concrete tends to hold daytime heat from the sun and buildings and businesses produce a great deal of heat. Water surfaces also tend to be much warmer than land on cold winter nights, and thus freezing is far less likely near large bodies of water such as Lake Washington, Puget Sound, or the Pacific Ocean.

Why Bridges Often Ice Up Before Other Roadways

The temperature of a road surface is affected by a number of factors, such as the amount of radiational cooling to space and the magnitude of heat coming up from the ground below. Heat conducted from below the road surface can lessen nighttime temperature falls and thus reduce the potential for icing. This is particularly true early in the winter season, after mild temperatures have warmed up the soil beneath the roadway. Bridges have air, a good insulator, beneath them and thus do not receive warmth from the ground below, yet they still radiate heat away to space in the same way as roads that are in contact with the ground. Thus, bridges are much more vulnerable to nighttime roadway icing compared to road surfaces in con-

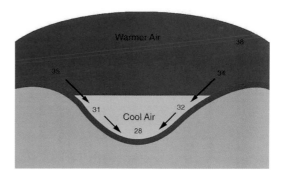

4.20. Cold, dense air tends to drain into low spots and valleys. Thus, the potential for ice on roadways is greatest in such areas.

tact with the ground, particularly early in the winter when the ground is relatively warm.

Why Ice Can Form on Roads Even When Air Temperature Is Above Freezing

Official temperature measurements are generally taken with thermometers in a sheltered enclosure approximately 6 feet above the ground, usually above a vegetated surface. So when a TV weathercaster gives the temperature at a nearby airport, that temperature is in fact the air temperature several feet above the surface, which can be *very* different than the air temperature immediately above the ground or the ground temperature itself. As mentioned earlier in this chapter, on clear nights when winds are light, the surface radiates heat to space much more effectively than the air above. During such conditions, temperature at ground level can be 2–5 °F cooler than the air temperature only a few feet above. Thus, frost can occur on a road even when official temperature observations report temperatures above freezing. During the day, the opposite situation can occur, with the road surface several degrees warmer than the air temperatures a few feet above. Air temperature readings available on many trucks and cars are similarly problematic because they measure temperatures several feet above the ground and thus show temperatures that are often warmer (night) or colder (day) than the road surface. The moral of the story is that at night motorists must be wary of icing when air temperatures drop below around 37 °F.

5

WINDSTORMS

||||||||||||||||||||||||||||||||||||||||||||||||||||||||||||||||||||||||||||||||||||||||||||||||||||||||||||||||||||||||||||||||||||||||||||||||||||

ALTHOUGH THE PACIFIC NORTHWEST ESCAPES THE THREAT OF HURRICANES AND

powerful tornadoes due to the cool water of the Pacific Ocean, the region is no stranger to

strong, damaging winds. Several major Pacific storms make landfall on the Northwest and

British Columbia coasts each winter, resulting in 40- to 70-mile-per-hour winds strong

enough to produce power failures and damage west of the Cascades. Less frequently, perhaps

two or three times a decade, windstorms of considerably greater magnitude occur: winds gust

above 70 miles per hour, resulting in extensive power failures that affect hundreds of thou-

sands of homes and generate tens or hundreds of millions of dollars of damage. Even less

frequently, roughly once every thirty years or so, the region experiences devastating storms with hurricane-force winds and catastrophic dam-age. Along the coast, winds exceeding 100 mph, and occasionally 150 mph, have accompanied these major events, particularly on exposed head-lands such as Cape Blanco on the Oregon coast and North Head just north of the Columbia River. In addition to windstorms produced by large Pacific storms, there are also more localized high-wind events associated with terrain features such as gaps and large barriers (see chapter 7).

The large storms that strike the Northwest during the fall and winter are known as *mid-latitude cyclones*. Like *tropical cyclones*, these storms are associated with low-pressure centers and winds that rotate counterclockwise in the Northern Hemisphere (clockwise in the Southern Hemisphere). But our storms are very different ani-mals than their tropical cousins. The energy source for midlatitude cyclones is the large horizontal variation of temperature found in the midlatitudes (30–60° north, 30–60° south latitudes), which lie between the warm tropics and the cooler, arctic regions. In contrast, tropical cyclones, including their most powerful members—hurricanes and typhoons—depend on warm water and large sup-plies of water vapor for their "fuel." In fact, tropi-cal cyclones require water temperatures of roughly 80 °F or more to gain sufficient heat and moisture for development. Since sea-surface temperatures of the eastern Pacific west of the Northwest rarely climb above the mid-50s, tropical storms have never visited our shores. On the other hand, mid-latitude cyclones are typically larger and can main-tain their strength over land far more effectively than tropical storms, which weaken rapidly when they are cut off from warm water.

An important measure of the strength of mid-latitude cyclones is the lowest sea-level pressure found in the center of the storm. As described later in this chapter, wind is produced by horizontal pressure differences, and lower central pressures generally result in larger pressure changes and stronger winds. Sea-level pressure is generally given in either millibars or inches of mercury.[1] A typical wintertime low-pressure center may have a central pressure of 990 millibars (29.23 inches), while the great Northwest windstorms have low centers as deep as 950 millibars (28.06 inches). Such low pressures are typical of weak to moderate hurricanes, but far exceed the central pressures of the most intense tropical storms, where sea-level pressure can drop to approximately 900 millibars (26.58 inches).

The Pacific Northwest is particularly vulnerable to strong windstorms due to its unique vegeta-tion and climate. The Northwest's tall trees, many reaching 100–150 feet, are force multipliers for regional windstorms, with much of the damage to buildings and power lines not associated with direct wind damage to structures, but with fall-ing trees (figure 5.1). Heavy precipitation in the autumn, which quickly saturates Northwest soils by the end of November, also enhances the damage potential, since saturated soils lose their adhesion and thus their ability to hold tree roots. Several species of Northwest trees, such as the western hemlock, are shallow-rooted and susceptible to uprooting during periods of wet soils and strong winds. Even the Douglas fir, which develops deep

---

1   Millibars are also known as hectopascals (hPA). The conver-sion between the units is 33.86 millibars = 1 inch of mercury. Average pressure at sea level is 1,013.25 millibars, or 29.92 inches.

root systems in thick, fertile soils, can be vulnerable to saturated soils and wind because it develops
shallow roots in poorly drained or shallow soils.

In describing wind events, meteorologists often
use the terms *sustained winds* and *gusts*. Wind
always varies in time, with lulls of weaker winds
followed by several seconds of greater speed. At
weather-observing sites, the sensors measuring
wind speed are called *anemometers* and provide
continuous measurement of these varying winds.
Averaging the instantaneous winds over several
minutes (typically two minutes) provides the sustained wind, while the highest short-term wind

**5.1. Most windstorm damage in the Northwest results from
fallen trees. The home of Professor Robert Houze in north
Seattle was severely damaged after the relatively weak
windstorm of March 30, 1997. Photo courtesy of Professor
Dale Durran, University of Washington.**

(typically a three- to five-second average) during
the observing period is known as the gust. Gusts
are typically much larger than sustained winds,
often by 25 to 50 percent, and are usually associated with the greatest damage.

Why are there wind gusts? Wind typically
increases with height in the atmosphere, since

air is slowed at low levels by the relatively rough surface, with its hills, trees, buildings, and other obstacles. Often the lower atmosphere is turbulent, with air moving up and down. During windy fall days this turbulence is often visible as leaves, papers, and other debris are tossed up and down by the atmospheric motions. Since winds generally increase with height, turbulent downward motions often bring down stronger winds—or gusts. Over water, approaching wind gusts are often indicated by darker, disturbed areas on the water surface (see figure 13.15). Strong gusts can be announced by an approaching roar, which is often heard several seconds before the gust's arrival.

5.2. Northwest Indian graphic depicting the giant thunderbird carrying off a killer whale. Graphic created by Professor Marvin Oliver, University of Washington Indian Studies Program, and provided courtesy of the Department of Atmospheric Sciences, University of Washington.

PACIFIC NORTHWEST HISTORICAL WINDSTORMS

Native American legends recognized the occurrence of strong winds and provided explanations of their origins. For example, the Quileute Tribe of the western Olympic Peninsula told stories about Thunderbird, a giant bird with wings twice as long as a war canoe and talons the size of oars. Great winds were produced as it flapped its huge wings, searching for its prey, the killer whale (figure 5.2). The Quileutes, like other coastal tribes, were aware that winter windstorms were more intense near the coast and generally moved to more protected inland camps during the stormy winter months.

European explorers of the region noted signs of the great southerly winds in the many tree falls to the north. For example, in June 1788 while sailing south of Cape Flattery, on the northwest tip of the Olympic Peninsula, the British trader John Meares was not only impressed by the great forests of the Northwest, but made the following report: "The

force of southerly storms was evident to every eye; large and extensive woods being laid flat by their power, the branches forming one long line to the North West, intermingled with roots of innumerable trees, which have been torn from their beds and helped to mark the furious course of their tempests."[2]

As non-native American settlers moved into the Northwest during the later half of the nineteenth century, they learned that Northwest windstorms were a force to be reckoned with. For example, in his memoir *Pioneer Days on Puget Sound*, written in 1888, Seattle founder Arthur Denny noted that "the heaviest windstorm since the settlement of the country" occurred on November 16, 1875. Denny described the storm as "a strong gale, which threw

---

2   Quoted in Robert E. Ficken, *The Forested Land: A History of Lumbering in Western Washington* (Seattle: University of Washington Press, 1987), 9.

down considerable timber and overturned light structures, such as sheds and outbuildings."[3]

January 9, 1880

An even stronger storm struck the region on January 9, 1880. Regarded by Portland's *Oregonian* as "the most violent storm . . . since its occupation by white men," the cyclone swept through northern Oregon and southern Washington, toppling thousands of trees, many 5–8 feet in diameter. Sustained winds of 60 mph began in Portland during the early afternoon, demolishing and unroofing many buildings, uprooting trees, felling telegraph wires, and killing one person. Scores of structures throughout the Willamette Valley were destroyed and hundreds more, including large public buildings, were damaged. Part of the roof of the Oregon State capitol in Salem was blown off, allowing snow to accumulate inside the building. Rail traffic was halted in most of northwest Oregon, virtually all fences in the Willamette Valley that were aligned east-west were downed, and every barn near the coastal town of Newport was destroyed. Wind gusts on the coast were estimated to have reached 138 mph. At Coos Bay, a three-masted schooner dragged its anchor, was blown onto the beach, and broke in two.

The description of the storm by the *Portland Oregonian* on January 10 suggested a near apocalyptic scene:

The most violent storm of wind which has visited this region since its occupation by white men occurred yesterday between the hours of 11 A.M. and 2:30 P.M. Not even among the traditions of the native Indian inhabitants of the country is there a record of a tempest so wild and furious in its aspect or so disastrous and terrible in its results . . . at 11 o'clock [it] increased to a storm, sweeping in a general direct course, but exhibiting whirls and eddies similar to the irregular movements of impetuous torrents of water. . . . By 12 o'clock the wind had reached a velocity of about fifty miles per hour, with occasional spurts at the extraordinary rate of sixty miles. The scene at this time, and for the succeeding two hours, was grand and terrible. The creaking of signs and buildings, the crash of falling awnings, the rumbling of tin roofs, the whistling chimes of electric wires, and above all and louder than all the fierce rage and roar of the tornado, united in a fearful and terrifying chorus. Men hurried hither and thither, eager, uncertain and fearful, women with white scared faces peered from the windows of their homes, dreading to remain yet knowing not whether to fly for safety, little children from the schools, which were soon dismissed, ran homeward with frightened haste, horses snorted in helpless fear, and even the dogs were affected with the universal terror.

January 29, 1921

The Great Olympic Blowdown of January 29, 1921, produced hurricane-force winds[4] along the northern Oregon and Washington coastlines and an extraordinary loss of timber on the Olympic

---

3 Arthur Denny, *Pioneer Days on Puget Sound* (1888; repr., Seattle: The Alice Harriman Co., 1908), available at the Washington State Library, www.secstate.wa.gov/history/publications.aspx.

4 Hurricanes are tropical cyclones with sustained winds reaching or exceeding 74 mph.

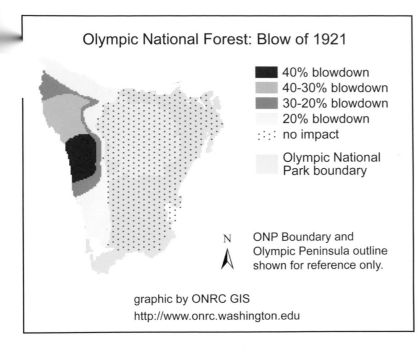

Olympic National Forest: Blow of 1921

40% blowdown
40-30% blowdown
30-20% blowdown
20% blowdown
:∷: no impact

Olympic National
Park boundary

N ⋀ ONP Boundary and
Olympic Peninsula outline
shown for reference only.

graphic by ONRC GIS
http://www.onrc.washington.edu

5.3. The windstorm of January 1921 produced extensive loss of timber along the coast of the Olympic Peninsula, but only minor damage in the interior of western Washington. Graphic created by the University of Washington Olympic Natural Resources Center using data provided by Olympic National Park.

Peninsula. As shown in figure 5.3, over 40 percent of the trees were blown down over the southwest flanks of the Olympic Mountains, with at least a 20 percent loss along the entire Olympic coastline. Billions of board feet of timber were uprooted or otherwise thrown down, much of it in remote regions that made salvage impossible. An official report at the North Head Lighthouse, on the north side of the mouth of the Columbia River, indicated a sustained wind of 113 mph,[5] with estimated gusts of 150 mph before the anemometer was carried away by the wind (figure 5.4). Although the coastal bluff seaward of the North Head site may have accelerated the winds more than those occurring over the nearby Pacific, the huge loss of timber around the lighthouse and along the Washington coast, eight times more than felled by the eruption of Mount Saint Helens in 1980, is consistent with a singular event.

The report of a U.S. Weather Bureau observer at the North Head station reveals the ferocity and sudden onset of the event. After pressure had fallen to its lowest level that day and wind speed had dropped temporarily below 30 mph, he and another staff member, Mrs. Hill, decided to travel to the nearby town of Ilwaco for supplies and mail. On the return trip from Ilwaco to North Head the errand became a life-threatening adventure:

The road from Ilwaco to North Head is through a heavy forest of spruce and hemlock for some distance. On the return trip, shortly before reaching the heavy timber, the wind came with quite a heavy gust. We saw the

5   This sustained-wind report actually recorded the "fastest mile," the wind speed associated with the period in which a mile of air moved past the North Head anemometer most rapidly. A fastest-mile wind speed of 113 mph implies an averaging period of approximately 30 seconds.

top of a rotted tree break off and fall out of sight in the brush. . . . We proceeded very slowly and with great care, passing over some large limbs that had fallen and through showers of spruce and hemlock twigs and small limbs blown from the trees. We soon came to a telephone pole across the roadway and brought the car to a stop, for a short distance beyond the pole an immense spruce tree lay across the road. We left the machine and started to run down the road toward a space in the forest where the timber was lighter. Just after leaving the car, I chanced to look up and saw a limb sailing through the air toward us; I caught Mrs. Hill by the hand and we ran; an instant later the limb, which was about 12 inches in diameter, crashed where we had stood. In three or four minutes we had climbed over two immense tree trunks and reached the place in which I thought was our only chance to escape serious injury or possibly death. The southeast wind roared through the forest, the falling trees crashed to the ground in every direction from where we stood. Many were broken off where their diameter was as much as 4 feet. A giant spruce fell across the roadway burying itself through the planks within 10 feet of where we stood. Treetops broke off and sailed through the air, some of the trees fell with a crash, others toppled over slowly as their roots were torn from the earth. In a few minutes there were but two trees standing that were dangerous to us and we watched every movement of their large trunks and comparatively small tops.

Between 3:45 and 3:50 p.m., the wind shifted to the south and the velocity decreased to probably 100 miles per hour or it may have been as low as 90 miles per hour. Shortly after 3:50 p.m. we started toward North Head. We climbed over some of the fallen trunks, crawled under others, and pushed our way through tangled masses of tops that lined the roadway. We supposed that all the houses at North Head had been lev-

eled and the wireless station demolished for we knew that the storm was the most severe that had occurred in the vicinity of the mouth of the Columbia for the last 200 years.[6]

(a)

(b)

5.4. (a) The lighthouse at North Head, Washington. On January 29, 1921, sustained winds reached 113 mph with gusts estimated to 150 mph at this site, the strongest winds ever observed over Washington State. Extraordinary damage occurred over much of coastal Washington during this event. Photo courtesy of Kathleen McCrory. (b) The North Head lighthouse sits on top of a coastal headland, which accelerates winds above those experienced over the ocean. Photo courtesy of Debbie Stika.

6   This account, as well as other information on the 1921 storm, can be found in two short articles in the *Monthly Weather Review* (January 1921): 34, 37.

It was estimated that 80 percent of the mature timber near North Head was razed during this storm. In addition, the wireless tower was demolished and all roofs in the vicinity were lifted from their structures. At the nearby town of Ilwaco, dozens of boats were torn from their moorings and dashed to pieces on beach bulkheads. Nearly all roads in the area were impassable. The storm took particular vengeance on local bird life, with a canvasback duck blown through a plate-glass window and Ilwaco chicken farmers finding their charges blown miles away. At Astoria, Oregon, on the south side of the Columbia, there were unofficial reports of gusts to 130 mph, while at Tatoosh Island, located at the Northwest tip of Washington, the winds reached 110 mph. Very strong, but lesser, winds were observed over Oregon's Willamette Valley and Puget Sound, with maximum gusts ranging

from 50 to 60 mph. In Seattle's Elliott Bay twenty-one barges broke their mooring lines and were driven into Puget Sound, while on land a number of Seattle greenhouses were destroyed and several dozen fires were ignited as a result of the strong winds. Power lines and telephone lines were downed across western Washington.

October 21, 1934

A more typical large windstorm, one that affected both the coast and western interior, occurred a little over a decade later (figure 5.5). The 1934 storm brought gusts of 60-70 mph to the interior of west-

5.5. The 1934 windstorm produced extensive damage around Seattle and lashed the harbor with unusually high waves and strong winds. Image courtesy of the *Seattle Times*.

ern Oregon and Washington, with higher winds on the coast, including an 87 mph gust at North Head. The storm removed roofs, overturned fishing boats, and at Boeing Field lifted a hangar off the ground that fell upon and destroyed four aircraft. Large swaths of forest were downed, and waves on Puget Sound and in the Strait of Juan de Fuca reached extraordinary heights of 20 feet. Five Seattle fishermen drowned when their boat, the *Agnes*, floundered in heavy seas near Port Townsend. On the Seattle waterfront, the Pacific liner *President Madison* became unmoored, hit and sunk two other ships, and then smashed into a dock before coming to rest. The smokestack of the central heating plant at the Church of the Immaculate Conception in Seattle toppled and crashed through the dome of the sanctuary, from which parishioners had left only ten minutes before. In addition to the loss of power and telephone lines throughout western Washington, the winds caused numerous fires throughout the region, making it the busiest day in the history of the Seattle Fire Department until that time. Large display signs were ripped from buildings across the city, and dozens of structures collapsed as a result of the strong winds. Twenty-two people in Washington and Oregon lost their lives.

The 1934 windstorm was associated with a deep low-pressure center that was located about 800 miles off of Eureka, California, the previous day; subsequently, it moved rapidly toward the northeast, making landfall over southern Vancouver Island. At Tatoosh Island, along the northwest tip of the Olympic Peninsula, sea-level pressure dropped to 977 millibars (28.85 inches) as the storm passed to the west.

November 3-4,1958

The 1958 event was rather unusual in that both strong southerly *and* northerly winds were observed over western Washington. An intense Pacific storm crossed the southwest Washington coast late in the afternoon of November 3, passing near Olympia and Mount Rainier during the evening hours. Extraordinary *northerly* winds of 45–65 mph were reported over Puget Sound north of Tacoma, while southwesterly winds of 70–90 mph were observed south of Olympia. A peak gust of 161 mph was reported at the Naselle air-defense-radar site on a 2,000-foot peak near the southwest Washington coast. Virtually every major highway along the coast of Oregon was blocked by fallen trees at some point during this storm. Astoria had a maximum gust of 75 mph, the Columbia River lightship reached 90 mph, and the Mount Hebo Air Force Radar Station, at an elevation of 3,174 feet in Oregon's coastal mountains, recorded several gusts of 130 mph, the maximum wind speed measurable on the anemometer at that location.

October 12, 1962: The Columbus Day Storm

By all accounts, the Columbus Day Storm was the most damaging windstorm to strike the Pacific Northwest since the arrival of European settlers. It may, in fact, be the most powerful nontropical storm to strike the continental United States during the past century. An extensive area stretching from northern California to southern British Columbia experienced hurricane-force winds, massive tree falls, and power outages (figure 5.6). In Oregon and Washington, 46 people died and 317 required hospitalization. Fifteen billion board feet of timber

5.6. The tower on Campbell Hall of Western Oregon State College collapsed during the height of the Columbus Day windstorm. Photo by Wes Luchau.

(more than a year's annual cut in these two states) were downed, 53,000 homes were damaged, thousands of utility poles were toppled, part of the roof of Portland's Multnomah stadium was torn off, and the twin 520-foot steel towers that carried the main power lines of that city were crumpled. At the height of the storm approximately one million homes were without power in the two states, with total damage conservatively estimated at a quarter of a billion (1962) dollars.

The Columbus Day Storm began east of the Philippines as a tropical storm—Typhoon Freda. As it moved northeastward into the mid-Pacific during October 8-10, the storm evolved into a midlatitude cyclone. During this process, known as *extratropical transition*, the cyclone changed structurally and switched energy sources, from the energy of the warm tropical ocean to that derived from the large temperature differences between the tropics and northern latitudes. Twelve hundred miles west of Los Angeles, the storm abruptly turned northward and began to deepen rapidly, reaching its lowest pressure (955 millibars, 28.20 inches) approximately 300 miles southwest of Brookings, Oregon, at around 7:00 AM on October 12 (see figure 5.20 for the storm track). Maintaining its intensity, the cyclone paralleled the coast for the next twelve hours, reached the Columbia River at approximately 5:00 PM with a central pressure of 956 millibars (28.23 inches), and crossed the northwestern tip of the Olympic Peninsula just before midnight. At most locations, the strongest winds followed the passage of an occluded front that extended southeastward from the storm's low-pressure center.

Over coastal regions and the offshore waters, winds gusted to well over 125 mph, and 60 to 120 mph gusts savaged the western interiors of Oregon and Washington. At the Cape Blanco Loran Station, sustained winds were estimated to reach 150 mph, with gusts to 179 mph. A 131 mph gust was observed at Oregon's Mount Hebo Air Force station, and gusts at the Naselle radar site in the coastal mountains of southwest Washington reached 160 mph. The winds at these three locations were undoubtedly enhanced by local terrain features, but clearly were extraordinary.

Away from the coast, winds gusted to 90 mph in Salem, Oregon, 116 mph at Portland's Morrison Street Bridge, 89 mph at Toledo, Washington, 100 mph at Renton, Washington, 83 mph at West Point in Seattle, 80 mph at Paine Field in Everett, Washington, 80 mph at Whidbey Island Naval Air Station, and 113 mph in Bellingham, Washington. Even in California fierce winds were observed, with sustained winds of 68 mph at Red Bluff, in the Central Valley, and gusts of 120 mph at Mount Tamalpais, just north of San Francisco. As is characteristic of most Northwest windstorms, both the storm and its associated winds weakened rapidly after landfall as the low center moved into British Columbia.

With sustained winds at some coastal locations exceeding 110 mph, the Columbus Day Storm was equivalent to a category 3 hurricane on the famous Saffir-Simpson scale, similar to Hurricane Rita that struck the southeast United States in September 2005. But the size of the Columbus Day Storm, like most midlatitude cyclones, far exceeded Rita or other hurricanes, and its path paralleled the coast, thus devastating a huge area stretching from central California to southern British Columbia. Striking today, the storm would have produced damage in the billions, if not tens of billions of dollars.

February 13, 1979: The Hood Canal Storm

Due to its track farther offshore, this storm caused less widespread damage over the Northwest than most events reviewed in this chapter. However, the interaction of strong southwesterly coastal winds with the high terrain of the Olympic Peninsula produced a small, but intense, low-pressure area to the northeast of the Olympics that accelerated southerly winds near the Hood Canal to over 100 mph. The result was the loss of the 3,200-foot western section of the Hood Canal floating bridge (figure 5.7) and a massive blowdown of timber, with over 80 percent of the trees toppled in some forests within 10 or 20 miles the bridge. For example, on the nearby Pope and Talbot tree farm, 26 million board feet were blown down, four to five times the amount that fell in that area during the Columbus Day Storm of 1962—the most destructive Northwest storm of modern times.

During the day preceding landfall, the cyclone center headed northeastward off the California and Oregon coasts, reaching its lowest pressure (965 millibars, 28.5 inches) at 4:00 PM on February 12, when the storm was roughly 400 miles west of the Oregon coast. Weakening as it continued to the northeast, the storm made landfall over north-central Vancouver Island at approximately 4:00 AM on February 13. Despite an overall decrease in the storm's intensity, an arc-shaped region of strong southerly winds persisted to the south of the low center and reached the Washington coast about this time.

5.7. Winds of more than 100 mph resulted in the failure of the western section of the Hood Canal Bridge on February 13, 1979. Photo courtesy of Tom Thompson and the *Peninsula Daily News*.

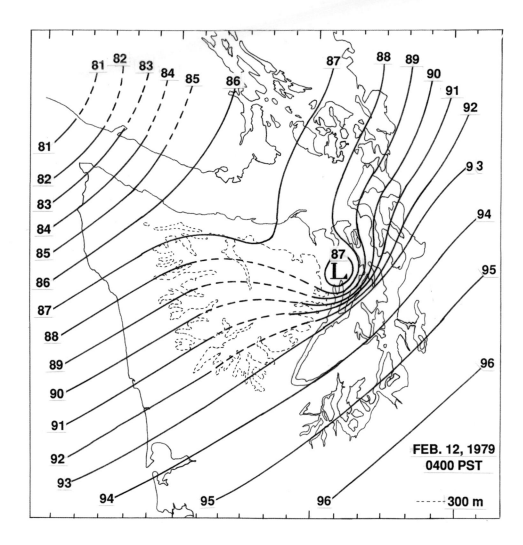

FEB. 12, 1979
0400 PST

----- 300 m

Over most of western Washington, the wind speeds associated with this event were not extraordinary for a winter windstorm. Maximum sustained winds at most locations were around 40 mph, with gusts under 75 mph. On the coast, the winds were somewhat higher, with the Cape Flattery Coast Guard Station reporting a maximum sustained wind of 56 mph and a peak gust of 98 mph. Astoria, Oregon, had a sustained wind of 54 mph, and five crewmembers of a 15,000-ton ship were washed to their deaths off Coos Bay, Oregon.

A detailed study of the storm by Professor Richard Reed of the University of Washington (Reed 1980) showed that the extraordinarily high

5.8. Sea-level pressure analysis at 4:00 AM PST on February 13, 1979. The lines are isobars (lines of constant pressure, where "87" indicates 987 millibars). An intense small-scale low-pressure center is apparent to the northeast of the Olympics. The large pressure variation south of the low accelerated winds to more than 100 mph over Hood Canal. Graphic from Reed (1980), courtesy of the American Meteorological Society.

winds near the Hood Canal Bridge were associated with a small but intense low-pressure center to the northeast of the Olympic Mountains (figure 5.8). Just as water passing over a large rock in a stream plunges downward after it passes the rock, with associated vortices and eddies immediately downstream of the rock, a similar phenomenon can occur in the vicinity of a large mountain barrier when the conditions are right. Specifically, when the winds approaching the crest of the Olympics are strong enough and the vertical stability of the air (the tendency for the air to return to its initial altitude when displaced vertically) is relatively low, an intense but small-scale low-pressure center, 10–20 miles in width, can develop downstream of the barrier. From theoretical studies, it appears that the conditions on February 13, 1979, were ideal for the formation of such a leeside low.

Air increases in speed as it moves from higher to lower pressure. Thus, already strong winds (30–40 mph) over southern Puget Sound accelerated rapidly over the Hood Canal area as they approached the small low-pressure area, which was centered near Port Townsend. Although the anemometer on the bridge was eventually lost, bridge tenders reported that wind gusts repeatedly reached their instrument's maximum reading (100 mph) before they were forced to abandon their station. The tenders also noted that sustained winds of 80 mph were observed for the two hours before the bridge failed. The Hood Canal Bridge was later replaced at a cost of over 140 million dollars.

The Hood Canal Storm was also associated with lesser but damaging winds near Enumclaw, a town near the western foothills of the Cascades. Enumclaw is downstream of a "weakness" or gap in the Cascades, and when low-pressure centers approach the coast, this town and its environs often experience strong easterly winds that accelerate through the gap and then descend the Cascade slopes (see chapter 8 for more detail on Enumclaw winds). With high pressure to the east of the Cascades and the low-pressure center of the Hood Canal storm offshore, easterly winds gusting above 65 mph struck Enumclaw prior to the strong southerly winds of the more general windstorm.

November 13-14, 1981

A number of Northwest windstorms have come in pairs or even triplets, as in this event of back-to-back storms. Two major storms traversed the Northwest during a forty-eight-hour period, with the first producing the most serious damage. The initial low-pressure center followed a similar course to that of the Columbus Day Storm, except that it tracked about 90 miles farther offshore, with landfall on central Vancouver Island. Over the eastern Pacific, this storm increased in intensity at an extraordinary rate, with the pressure dropping by approximately 50 millibars (1.48 inches) during the twenty-four-hour period ending 4:00 PM Pacific standard time on November 13 (figure 5.9). At its peak over the eastern Pacific, the storm attained a central pressure of just under 950 millibars (28.05 inches), making it one of the most intense Northwest storms of the century; two ships just offshore reported gusts of 100 mph. The second, but weaker, storm followed a nearly identical path a day later, with a central pressure reaching only 988 millibars (29.17 inches).

Over land, the maximum gusts from the first storm ranged from approximately 100 mph over coastal Oregon to nearly 70 mph over the interiors of western Oregon and Washington. Winds exceeding 50 mph spread into coastal northern California after 4:00 PM on November 13. Subsequently, Brookings, Oregon, experienced sustained winds of 82 mph, followed by 92 mph gusts at North Bend, and 97 mph gusts at Coos Bay. At the Coast

Guard station in North Bend there was an unofficial report of a gust exceeding 120 mph, and a ship 250 miles off the southern Oregon coast reported 33-foot seas. By the early morning of November 14, strong winds hit the Willamette Valley and the northern Oregon coast, and a few hours later winds rapidly increased over western Washington. At Seattle's Evergreen Point Bridge, sustained winds of 45 mph with gusts to 75 mph resulted in nearly four hundred thousand dollars in damage and an eleven-hour closure. Maximum gusts reached 67 mph at Seattle-Tacoma Airport. By the late afternoon of November 14, the winds were dying down over western Washington, only to accelerate again during the afternoon of the 15th as the second, but generally weaker, storm tracked offshore. During

the second storm the winds gusted in Oregon to 62 mph at Medford, 92 mph at North Bend, 75 mph at Astoria, 71 mph at Salem and Portland, and in Washington to 52 mph at Seattle-Tacoma Airport, 79 mph at the Evergreen Point Bridge, and 66 mph at Camano Island. Paradoxically, the second weaker storm did more damage to trees, perhaps because the first storm had already weakened or partially dislodged tree roots.

Thirteen fatalities were directly related to the November 1981 storms: five in western Washington and eight in Oregon. Most were from falling trees, but four died in Coos Bay during the first storm when a Coast Guard helicopter crashed while searching for a fishing vessel that had encountered 30-foot waves and 80 mph winds. The fishermen were never found. Although these events are reminiscent of the losses during the Halloween 1991 "perfect storm" of movie fame, the Northwest storm was far stronger than the East Coast system—perhaps Northwest "perfect storms" are more perfect than their eastern cousins. The stronger, Northwest 1981 storm resulted in massive power outages, with four hundred thousand Puget Power customers losing electricity. In downtown Seattle, a 70-foot cable-relay tower was blown down, while across the Sound in Bremerton winds blew the USS *Oriskey* from its moorage in the Puget Sound Naval Shipyard. Every Washington State ferry run was cancelled and all major western Washington bridges were closed.

As in many great windstorms, in the wake of the first November 1981 cyclone a large and abrupt increase in pressure occurred as strong winds spread over the interiors of western Oregon and Washington. At some locations the pressure rose over 10 millibars (0.295 inch) in three hours. With pressure falling ahead of the low-pressure center and rapidly rising behind, extremely large horizontal pressure differences developed over the region that helped to drive the powerful winds.[7]

An interesting aspect of the November 1981 windstorms was the storm insurance policy held by Puget Power. The company's power grid had sus-

tained massive damage from the dual storms that totaled in the millions of dollars. They did have insurance, but with a deductible of one million dollars *per event*. The whole issue went to court, where a jury decided that it really was one "storm," even though there were two low-pressure centers. Clearly, the legal and meteorological definitions of a storm are not always the same.

The guidance by National Weather Service (NWS) computer models was nearly useless during the first event, with the twenty-four-hour forecasts providing little hint of intensification. The Seattle NWS office was able to make a timely (eight-hour lead time) forecast only because Harry Wappler, the chief meteorologist at KIRO-TV at the time, rushed over with a video tape containing the latest satellite animation, allowing NWS forecasters to extrapolate the future trajectory of the intense system (this was before NWS offices were equipped with weather-satellite animation capabilities).

January 20, 1993: The Inauguration Day Storm

Probably the third most damaging storm since 1960 (with the Columbus Day Storm being number one and the December 2006 storm in second place) struck the Northwest on the day of President Bill Clinton's inauguration, January 20, 1993. Winds of over 100 mph were observed at exposed sites in the coastal mountains and the Cascades, with speeds exceeding 80 mph along the coast and in the interior of western Washington. In Washington six people died, approximately 870,000 customers lost power, 79 homes and 4 apartment buildings were destroyed, 581 dwellings sustained major damage, and insured damage was estimated at 159 million dollars.

---

7   For example, there was a 20.3 millibars (0.6 inches) pressure difference between Salem and Bellingham at 9:30 AM on November 14.

The Inauguration Day Storm intensified rapidly during the day preceding landfall on the northern Washington coast. At 4:00 PM Pacific standard time on January 19, the low-pressure center was approximately 600 miles east of the northern California coast, with a central sea-level pressure of 990 millibars (29.22 inches). The storm then entered a period of rapid intensification, with the central pressure reaching its lowest value (976 millibars, 28.82 inches) at 7:00 AM on January 20, when it was located about 20 miles east of the Columbia River outlet (figure 5.10a). A secondary trough of low pressure extended south of the main low center, and within this trough the horizontal pressure differences and associated winds were very large (figure 5.10b). During the next six hours, as the low-pressure center passed west and north of the Puget Sound area, the secondary trough moved northeastward across northwest Oregon and western Washington, bringing hurricane-force winds and considerable destruction.

As the storm moved along the Oregon coast, winds gusted to 86 mph at Cape Blanco and 84 mph at Arch Cape, with unofficial reports of gusts in excess of 100 mph over northwest Oregon. These hurricane-force winds produced widespread power outages over coastal Oregon and the loss of millions of dollars of timber in the coastal mountains southeast of Astoria. As the storm moved northward along the Washington coast, winds gusted to 98 mph at Cape Disappointment at the western terminus of the Columbia River, 94 mph at the Hood Canal Bridge, 75 mph at Alki Point, 80 mph at Enumclaw near the Cascade foothills, and a record 88 mph on the roof of the atmospheric sciences building at the University of Washington in Seattle. The author has experienced the passage

of hurricanes along the East Coast, and the arrival of the Inauguration Day Storm at the University of Washington was every bit as intense. The 64 mph gust at Seattle-Tacoma Airport was the third-strongest in fifty years (the record being the 69 mph gust during the December 15, 2006, storm). Near the Cascade crest, winds exceeded 100 mph several times over a two-hour period at Stampede Pass and reached 116 mph at the Alpental ski area in Snoqualmie Pass. For the first time ever, both floating bridges across Lake Washington were closed, as was the Tacoma Narrows suspension bridge. The massive power outages accompanying the storm caused businesses and schools throughout western Washington to close midday; some schools sent children home when dangerous winds were still pummeling the region (a big mistake that could have had tragic results). A power outage north of Seattle in Edmonds resulted in nine million gallons of raw sewage inundating city streets, while near Toledo, Washington, Interstate 5 was closed when power lines were downed over four lanes.

Official National Weather Service forecasts were excellent for this storm, with the release of a high-wind watch at 1:30 PM Pacific standard time and a high-wind warning at 10 PM the day before (January 19). These exceptional predictions were generally ignored by the media, which were busy covering Clinton's inaugural. The skillful forecast of this event reflected, in part, the substantial improvement in numerical weather prediction that had occurred during the previous ten years. While nearly every major windstorm was poorly predicted prior to 1990, advances in observing technology and numerical weather prediction have resulted in consistently better forecasts of major storm systems during recent years.

(a)

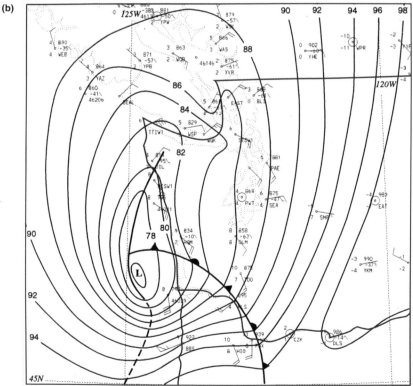

(b)

5.10. (a) Infrared satellite picture at 7:00 AM PST on January 20, 1993, of the Inauguration Day Storm, showing the low-pressure center (in the middle of the circular swirl of clouds) making landfall on the Washington coast. In infrared pictures, whiter clouds are higher, and darker areas are lower. Image from National Atmospheric Administration/National Weather Service GOES-West weather satellite. (b) A surface-pressure analysis of the Inauguration Day Storm at the same time. The lines are isobars (lines of constant pressure, in millibars). Note that the largest pressure changes (and strongest winds) are found south of the low-pressure center in the storm's "poisonous tail." Graphic from Steenburgh and Mass (1996), courtesy of the American Meteorological Society.

December 12, 1995

Of all the major windstorms to strike the Pacific Northwest, none was better forecast or more intensively studied than the event of December 12, 1995. Hurricane-force gusts and substantial damage covered an extraordinarily large area from San Francisco Bay to southern British Columbia, leaving five fatalities and over two hundred million dollars of damage in its wake. Early in the day, the storm struck northern California with gusts of 103 mph at San Francisco, 75 mph at Eureka, and 75 mph in Oakland, resulting in numerous fallen trees and three deaths. In Oregon, winds at Sea Lion Caves near Florence reached 119 mph before the anemometer failed, winds gusted to 86 mph at North Bend, and attained 107 mph at Newport (figure 5.11). Both Cape Blanco and Astoria had maximum winds in excess of 100 mph. Sea-level pressure at Astoria dropped to 965 millibars (28.51 inches), an all-time record low for that location. Winds within the Willamette Valley surpassed 60 mph at several sites and, with very wet soil from an unusually rainy fall, many large trees were uprooted.

Over western Washington sustained winds reached 30–50 mph, with gusts to 50-90 mph. North Bend and Seattle were buffeted by maximum gusts of 78 mph and 59 mph, respectively, with the latter location measuring its all-time record low pressure (970 millibars, 28.65 inches). Over the waters of Puget Sound the winds were even stronger than over land: a ship just outside of Elliott Bay reported sustained winds of 60–70 mph, with gusts to 90 mph; the ferry terminal at Mukilteo reported sustained winds of 60–70 mph, with a gust to 86 mph; and gusts reached 76 mph on the

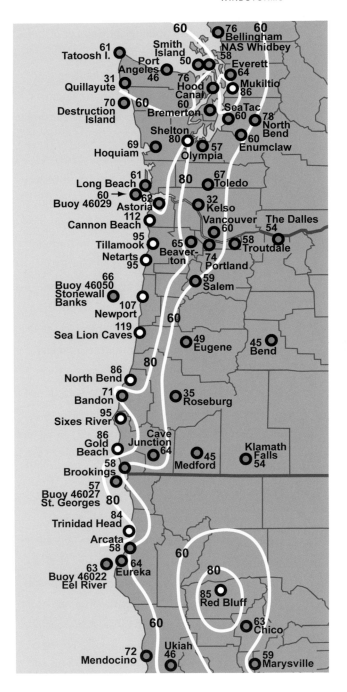

5.11. Maximum wind gusts (mph) during the December 12, 1995, windstorm. Winds exceeding 100 mph struck portions of the Oregon Coast, with 70–90 mph gusts at some inland locations. An analysis of the wind speeds is shown by the white lines (60 mph and 80 mph). White dots indicate winds greater than 80 mph. Graphic courtesy of Wolf Read.

Hood Canal Bridge. Approximately 400,000 homes lost power in western Washington, with nearly complete blackouts on Bainbridge, Vashon, and Mercer islands. To the south, 205,000 customers lost power in Oregon, while in northern California the total was 714,000.

The December 12, 1995, event was the most skillfully forecast windstorm in Northwest history. The computer weather models began predicting an intense event three to four days ahead of time, and the day before the storm struck the National Weather Service provided a strongly worded warning for a powerful, damaging windstorm. The media went wild, with television reporters reporting live from seemingly dozens of locations. Government, educational institutions, and businesses took the warnings seriously, protecting property and closing early. The forecasts released the morning of December 12 not only correctly predicted the storm strength, but provided timing that was accurate to within an hour.

As luck would have it, a major atmospheric observational program called COAST (Coastal Observation and Simulation with Topography Experiment) was underway during the December 1995 windstorm, and a National Oceanic and Atmospheric Administration (NOAA) research aircraft, the P3, was used to study the structure of the storm both offshore and as it crossed the coastal mountains of Oregon and Washington. The P3, which is also used for hurricane reconnaissance, not only observes the weather along its flight path, but also has Doppler radars that describe the three-dimensional wind and precipitation fields around the plane for a distance of fifty to a hundred miles. Flying offshore of the Oregon coast at around 4,000 feet, the plane experienced winds of 100–120 mph in a highly turbulent environment, with salt spray reaching the plane's windshield as high as 2,000 feet above the wind-whipped seas.

February 7, 2002: The South Valley Surprise

Although the National Weather Service has had increasing success in providing timely and accurate predictions of major Northwest windstorms, failed forecasts still occur. One of the most dramatic prediction "busts" occurred on February 7, 2002, when sustained winds of 50–60 mph, with gusts reaching 70–90 mph, struck the southern Willamette Valley and the southern Oregon coast. In contrast to other recent windstorms, there was virtually no warning for this event—thus, the storm is often called the South Valley Surprise (figure 5.12).

This storm developed rapidly over the Pacific as it approached the southern Oregon coast during the morning of February 7 and made landfall between North Bend and Newport around 4:00 PM, passing south of Portland an hour or two later. Highest wind gusts on the coast ranged from 84 mph at Gold Beach, just north of the California-Oregon border, to 88 mph at Bandon, just south of Coos Bay, and 100 mph at Winchester Bay, 20 miles farther to the north. In the interior, the strongest winds were in the vicinity of Eugene, Oregon, where a sustained wind of 49 mph, with gusts to 70 mph, was observed. Damage was widespread, with thousands of trees and hundreds of power poles toppled, power outages throughout the area, and extensive damage to homes, roofs, and buildings. Perhaps the most dramatic example was the line of toppled poles, a quarter to a half-mile long, that were found near Eugene (figure 5.13). In

Back to the Capitol

A special session of the Oregon Legislature opens today to resolve a massive state budget shortfall.

CITY/REGION / 1C

Ducks come up short at Stanford

Men fall in OT; Oregon women also lose close one

SPORTS / 1D

Morning shower
High 52, Low 38
WEATHER / 18D

50 Cents

The Register-Guard

www.registerguard.com

Victory blossoms for rose grower

■ **Chase Gardens:** Oregon justices reinstate a verdict over a lien that put the 100-year-old company out of business.

By BILL BISHOP
The Register-Guard

After 11 years and six appeals, Chase Gardens won its David-vs.-Goliath legal battle against Northwest Natural Gas Co. with a ruling Thursday by the Oregon Supreme Court reinstating a $1.9 million jury verdict against the gas company.

In the decision, its second in the case, the state's highest court said a Lane County Circuit Court jury was right when it found in 1995 that the gas company breached its duty to deal in good faith by filing a lien against the rose grower for an unpaid gas bill. The move came a few weeks before Chase Gardens would have harvested its $500,000 Valentine's Day rose crop.

The lien caused Chase Gardens' bank to freeze its line of credit, prompting the 100-year-old family-owned business to lay off its 22 employees and close in January 1991.

In earlier appeals, legal decisions favored first one side, then the other. With the latest ruling, no further appeal is possible.

But the victory is bittersweet for Bruce and Katherine Chase. The settlement doesn't come close to covering the family's losses, they said.

"I'm not going to give up my day job," said Bruce Chase, 70, who owned the business with his wife and their three children. He has worked part time as a maintenance man since Chase Gardens closed.

"At this point, after 11 years, you're afraid to think it might be over and you might have won," said Katherine Chase, 68. "It's hard to get over. You tell yourself it's in the past, but it's not easy to do."

A Northwest Natural spokesman said the company's lawyers hadn't yet reviewed the decision and could offer no comment.

Before Chase Gardens closed and filed the lawsuit in 1991, it operated a cut flower business in 1 million square feet of greenhouses along Centennial Boulevard near Interstate 5.

The lawsuit claimed that the gas company acted unfairly when it placed a lien against the company over an unpaid $49,000 bill and demanded a $100,000 payment to remove the lien.

In past years, according to the suit, the gas company recognized the cyclical nature of the cut flower business and worked with the company as it struggled to keep up with payments.

Northwest Natural argued in court that it only did what any creditor is entitled to do: attempt to collect from a delinquent debtor.

In 1995, a jury awarded Chase Gardens $1.9 million in damages.

Turn to CHASE, Page 8A

STORMY WEATHER

Winds land sneak punch

THOMAS BOYD / The Register-Guard

Eugene firefighters work to free three men from a Chevrolet pickup truck on 13th Avenue after a large evergreen tree fell on the vehicle. Two of the men, Jesse Faunce and driver Matt Sprick, were rescued without injuries, but passenger Jake Clifton was transported to Sacred Heart Medical Center with undetermined injuries.

Surprise storm packing 70 mph gusts downs trees, damages property

By JEFF WRIGHT
The Register-Guard

A sneaker windstorm with gusts reaching 70 mph ambushed the southern Willamette Valley late Thursday afternoon, downing hundreds of trees, damaging houses and cars, snarling traffic, ripping off highway signs and leaving thousands of Lane County residents without power.

At least one man was seriously injured when a 50-foot spruce tree crashed onto two trucks at the corner of East 13th Avenue and Pearl Street in Eugene.

The falling trees smashed dozens of vehicles, homes and businesses, causing extensive property damage, though officials said it's too early to calculate a dollar figure. Several school districts around the area — from Junction City to McKenzie — called off classes today because of storm damage.

The storm — one of the most powerful in recent memory — began on the southern coast, then worked its way inland to Coos Bay and up the Willamette Valley.

Doug Putscher, Lane County's road maintenance manager, said he dispatched about 50 crew members to locations throughout the county.

"Dexter, Fall Creek and Cottage Grove have moderate to heavy damage, and Veneta is a war zone," Putscher said. "My

Turn to WINDS, Page 8A

UTILITY HELP

■ Eugene Water & Electric Board: 484-2300
■ Eugene Public Works: 682-4800
■ Springfield Utility Board: 746-8451
■ Emerald People's Utility District: 746-1583
■ Blachly-Lane Electric Co-op: 688-8711
■ Lane Electric Co-Op: 484-1151

TRAVEL

■ Call the state Department of Transportation's toll-free number, (800) 977-6368
■ Visit www.TripCheck.com, the Department of Transportation travel information Web site

SCHOOLS

■ Check the Internet at www.registerguard.com, or Lane Education Service District Web site at www.valleyinfo.net; tune to local radio stations

Eugene firefighters help Jesse Faunce out of the back of the Chevrolet pickup truck on Thursday night.

Three men freed after tree flattens their pickup truck

By SUSAN PALMER
The Register-Guard

Jesse Faunce was riding in the back of a four-seater Chevrolet pickup when Thursday's ferocious winds tore a 50-foot spruce tree from the ground and dropped it on the cab of the truck, smashing the roof almost level with the doors.

Faunce, Jake Clifton and driver Matt Sprick were hauling a snowmobile west on 13th Avenue near Pearl Street about 4:30 p.m. when the tree toppled in front of St. Mary's Episcopal Church, bringing down a live power line on top of the truck.

The tree was so tall that it landed on the Chevy in the right-hand turn lane and a Toyota pickup truck in the far left lane.

Passers-by were able to help the driver of the Toyota out of his truck unharmed, but the tree — 3 feet in diameter — rested solidly on top of the Chevy.

Firefighters spent the next 2½ hours trying to free the three men in the Chevy, hampered by the unstable tree and the live power line.

Turn to RESCUE, Page 8A

5.12. The front page of the *Eugene (OR) Register-Guard* documented the substantial damage and poor forecasts associated with the February 7, 2002, windstorm. Image courtesy of the *Register-Guard*.

5.13. Strong winds during the South Valley Surprise storm of February 7, 2002, toppled many power poles, including this line of poles on Colton Road near Eugene, Oregon. Photo courtesy of Wolf Read.

the southern Willamette Valley this was probably the most severe blow since the 1962 Columbus Day Storm.

Why was the National Weather Service unable to forecast this storm? One answer undoubtedly lies in the failure of the computer prediction models that day. Figure 5.14 shows a weather-satellite image of the February 7, 2002, storm at 4:00 PM, overlain with a pressure analysis for the same time based on observations (yellow lines). A low-pressure center was positioned over northwest Oregon, with a large difference in pressure and associated strong winds over southwest Oregon. Also shown is the forty-eight-hour prediction of the NWS Eta model, the highest resolution weather-prediction model available to NWS forecasters (blue lines). As the discrepancies between these two sets of lines reveal, the computer prediction completely missed the storm, and this was true of even shorter-range computer forecasts. Why did the computer models have so much trouble with this event? Although there is no definitive answer to this question, lack of sufficient observations over the ocean and deficiencies in the model's ability to use or "assimilate" available observations were probably

involved. The fact that the storm was relatively compact and developed so close to shore surely contributed as well. However, acknowledging these deficiencies, human failure is also indicated, since rapid pressure falls on the coast and an ominous signature of a rapidly developing storm in satellite pictures were apparent. Another problem is the absence of coastal weather-radar coverage for the Northwest, which prevents local meteorologists from viewing the inner details of approaching storms.

December 14–15, 2006: The Hanukkah Eve Storm

The most damaging winds since the Columbus Day Storm of 1962 assaulted the region on December 14–15, 2006. With winds gusting to 90 mph along the coast, 80 mph in the eastern Strait of Juan de Fuca, and 70 mph over the Puget Sound lowlands, over 1.3 million customers lost power in western Washington, at least thirteen individuals lost their lives, and early estimates of damage ranged from five hundred million to a billion dollars. Some of the strongest winds occurred at Tatoosh Island (78 mph) and Ocean Shores (73 mph) on the Washington coast, at Smith Island (76 mph) and Padilla Bay near Burlington (85 mph) in the eastern strait, and at Poulsbo (74 mph) and the Hood Canal Bridge (74 mph). In the Cascades, winds reached 100 mph at Sunrise on Mount Rainier and 113 mph at Chinook Pass. Oregon hardly escaped this powerful storm. A number of locations on the Oregon coast experienced gusts exceeding 90 mph (Newport, 106 mph; Garibaldi, 93 mph; Rockaway Beach, 97 mph), and a site on Mount Hebo in the Oregon coastal mountains topped 114 mph. Lesser but still powerful winds (50–80 mph gusts) hit the

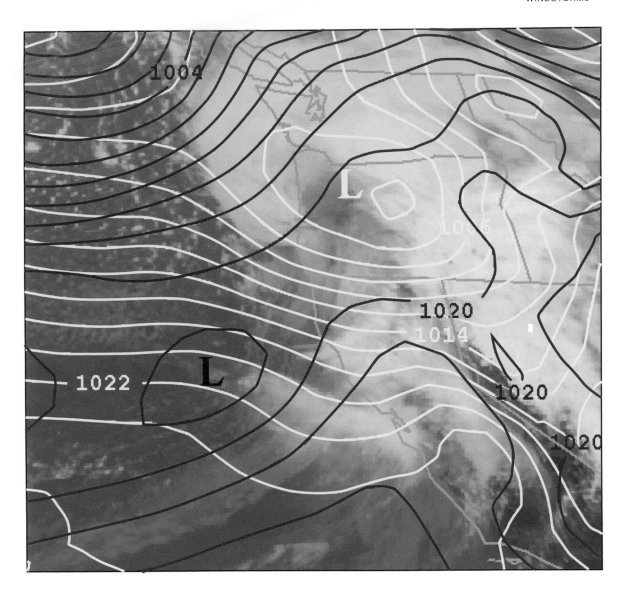

Willamette Valley. By the end of the event nearly 250,000 power customers in Oregon were in the dark, and the damage toll in Oregon was similar to the powerful December 12, 1995, event.

The December 2006 storm approached the region along a more westerly trajectory than is typical of major Northwest windstorms, which generally enter from the south to southwest (figure 5.20 shows the storm track). Intensifying as it approached the coast, the storm's central pressure

5.14. Satellite image of the February 7, 2002, windstorm at 4:00 PM PST. Also shown is a pressure analysis based on observations (yellow lines) and a forty-eight-hour pressure forecast by the National Weather Service Eta weather prediction model (blue lines). It is clear that the forecast of this storm was very poor. Image from National Oceanic and Atmospheric Administration/National Weather Service GOES-West weather satellite.

fell rapidly to approximately 973 millibars (28.70 inches) just prior to making landfall along the central coast of Vancouver Island. A satellite picture on the afternoon of December 14 reveals the swirling clouds around the low, cold unstable air

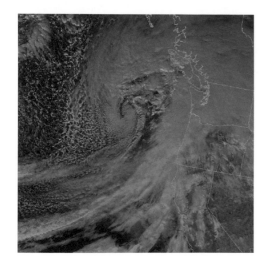

5.15. Visible satellite image at 1:00 PM on December 14, 2006. The low-pressure center is located at the center of the swirling clouds offshore of the Oregon-Washington border. Image from National Oceanic and Atmospheric Administration/National Weather Service GOES-West weather satellite.

behind it (marked by the speckled white clouds to its west), and a frontal cloud band that starts north of the system, crosses the Washington coast, and then heads southwest over the Pacific (figure 5.15). A highly realistic simulation of the storm just prior to landfall displays not only the low center, but also a region of large pressure change on the southern side of the storm (figure 5.16). Often called the "poisonous tail of the bent-back occlusion" by meteorologists, this region of large pressure difference south of the low-pressure center is characteristic of most oceanic low-pressure systems and is associated with the strongest winds, since winds are produced by differences in pressure. As the area of low pressure moved inland over southern British Columbia, the region of strongest pressure change and winds—the poisonous tail—moved right over western Washington, with disastrous results.

Over western Washington, the damage associated with this storm substantially exceeded that

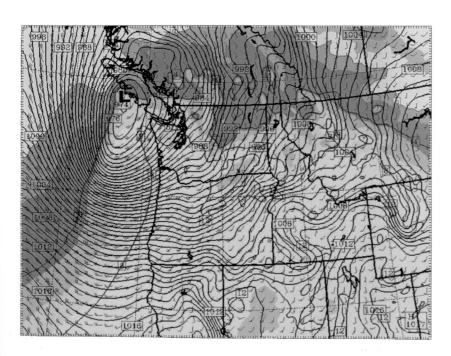

5.16. Three-hour forecast of sea-level pressure (solid lines, millibars) and near-surface air temperature (shading, °C) valid at 7:00 PM on December 15, 2006. Winds are indicated by the small wind flags, with the flag pointing in the direction of the wind. The low-pressure center was just making landfall on Vancouver Island. Note the area of large pressure change—and strong winds—to the south and west of the low center. This area of high winds subsequently moved northeastward over western Washington. Yellow areas represent warmer temperatures, while cooler air (blue and white) was found to the north and west of the low center. Figure from the University of Washington MM5 Forecast Model.

5.17. Scenes such as this on West Mercer Way of Mercer Island, Washington, were widespread over the Northwest in the aftermath of the December 2006 windstorm.

of the Inauguration Day Storm. Nearly double the number of customers lost power than in the 1993 event and restoration took several weeks for some neighborhoods. The scene shown in figure 5.17 of closed roads, fallen trees, and downed power lines was repeated over the region, with the most extensive damage in wooded eastern Puget Sound communities and Mercer Island. Some have claimed that the winds were stronger in 2006 than in 1993, but a close examination does not suggest this. As shown in table 5.1, over western Washington there is no clear winner, with the strongest winds nearly split between the storms.

If the winds of the 1993 and 2006 storms were of roughly equal magnitude, why did the later storm produce more tree fall and damage? Two answers come to mind: a significant difference in the precipitation during the weeks preceding each storm and the population increase of the intervening years, with much of the growth occurring in suburban and rural areas with large trees. The two-month period preceding the 2006 windstorm was extraordinarily wet over the entire Northwest. Many locations received 200 percent or more of normal rainfall and a number of observing sites broke their *all-time* precipitation records for November. Even more impressive, some locations, such as Seattle-Tacoma Airport, exceeded their

LOCATION	1993 WIND SPEED	2006 WIND SPEED
Tacoma	62	**69**
Renton	**74**	51
Seattle-Tacoma Airport	64	**69**
Boeing Field	**70**	56
West Point Lighthouse	60	**70**
Everett Paine Field	**67**	66
Smith Island	52	**76**
Bellingham	**59**	55

Table 5.1. Comparison of maximum wind gusts (mph) between the 1993 Inauguration Day Storm and the 2006 Hanukkah Eve event. Bold numbers indicate the stronger winds for each location.

all-time record precipitation for *any* month. On the day preceding the storm, heavy rain struck most locations west of the Cascades, with some places receiving as much as an inch of rain in a single hour, an amount that probably exceeded the all-time records for such a brief period. In short, the regional soils were completely saturated before the December 14–15 windstorm, with antecedent precipitation amounts that entered the record books. As noted earlier, saturated soils are less capable of holding tree roots, and thus trees are more vulnerable to toppling under such conditions.

Another contributor to increased damage in 2006 was surely the construction of homes in areas that were previously forested. Typically, builders clear sufficient land for a home and landscaping, but leave some tall trees on the property as a scenic backdrop. Such large openings in a forest provide entry for strong wind gusts that can topple trees that had previously enjoyed protection. Many of the newer developments on the east side of Puget Sound sustained substantial damage from such newly exposed trees.

December 2–3, 2007

One of the region's most unusual, but intense, windstorms struck the northern Oregon and southern Washington coasts for an extended period during December 2–3, 2007. Two-minute sustained winds of 50–75 mph, with gusts reaching 125 mph and more, produced extensive tree falls, building damage, and power outages from Lincoln City, on the central Oregon coast, to Grays Harbor County of Washington. The extraordinary winds toppled or snapped off trees, including extensive swaths of forests, a sign of an extreme event (figure 5.18). The December 2–3 storm was highly localized: while winds were blowing at hurricane force over the coastal zone, surface winds were light to moderate over Puget Sound and the Willamette Valley.

The December 2007 storm was singular in several ways. First, most major Northwest windstorms are associated with intense and fast-moving low-pressure centers that move northward up the coast. Such rapidly moving storms generally produce strong winds for only a few hours. In contrast, this windstorm was associated with a persistent area of large pressure differences, between a slow-moving low offshore and much higher pressure over the continent, that remained over the northern Oregon/southern Washington coastlines for nearly twenty-four hours (figure 5.19). Another unusual aspect of this storm was the associated heavy pre-

5.18. Hurricane-force winds with gusts reaching 125 mph struck the northern Oregon and southern Washington coasts during December 2–3, 2007. In exposed locations, swaths of trees were snapped off midsection from winds second only to those observed during the 1962 Columbus Day Storm. This picture shows a large area of broken trees above Oregon State Route 26, about 4 miles east of the US 101/State Route 26 junction. Photo courtesy of Wolf Read.

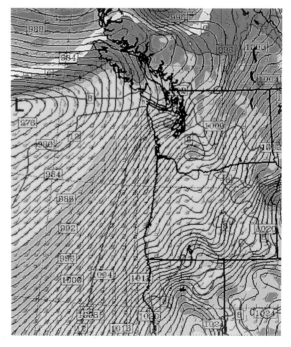

5.19. Short-term forecast of sea-level pressure, surface winds, and low-level temperature valid at 4:00 AM on December 3, 2007. The solid black lines are isobars (lines of constant pressure, in millibars) and winds are indicated by the small wind flags, with the flag pointing in the direction of the wind. Orange and yellow areas are associated with the warmest temperatures; white and blue regions are relatively cold. Note the large difference in pressure over the coastal zone between the offshore low and higher pressure over land. Such large differences in pressure are associated with strong winds. Image from the Department of Atmospheric Sciences, University of Washington.

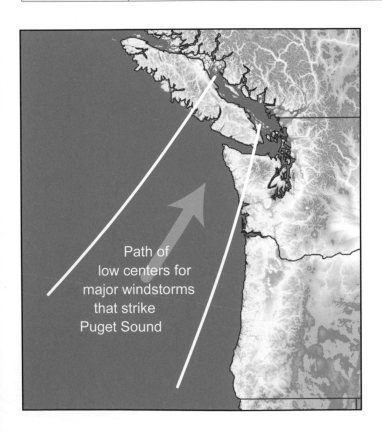

170° W   165° W   160° W   155° W   150° W   145° W   140° W   135° W   130° W   125° W   120° W

50° N

45° N

40° N

35° N

Path of
low centers for
major windstorms
that strike
Puget Sound

5.20. Tracks of some major Northwest windstorms.
Note that most move from the Pacific west of California
to the northeast over or just to the north of the Pacific
Northwest. Graphic courtesy of Bri Dotson.

5.21. Typical path of low-pressure centers that bring
strong southerly winds to the Puget Sound region.

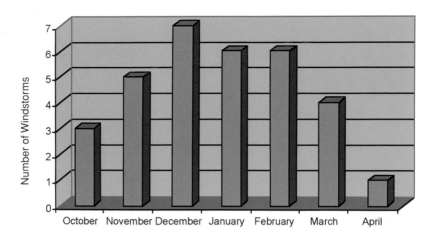

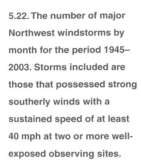

5.22. The number of major Northwest windstorms by month for the period 1945–2003. Storms included are those that possessed strong southerly winds with a sustained speed of at least 40 mph at two or more well-exposed observing sites.

cipitation over the west-side lowlands. Generally, fast-moving windstorms do not have time to produce large rainfall totals, and heavy rain generally precedes the strong winds. In this case, the heavy inland rainfall and strong coastal winds occurred simultaneously.

GENERAL WINDSTORM CHARACTERISTICS

Most of the great Northwest windstorms have followed similar tracks over the eastern Pacific. Starting over the central Pacific west of California, such storms usually move northeastward before swinging toward the north as they approach the West Coast (figure 5.20). The strongest winds from these storms are inevitably from the south, with the greatest speeds occurring when the low center passes north of the location in question. For the Puget Sound region, the most powerful southerly winds typically occur when a deep low-pressure center moves from southwest to northeast, crossing the coast between the northern tip of the Olympic Peninsula to central Vancouver Island (figure 5.21). In contrast, the

strongest winds hit Portland or the northern Willamette Valley when the low center moves across southern Washington State.

Strong Northwest windstorms have struck the region in the months of October through April, with the largest numbers occurring November through February (figure 5.22). Interestingly, the strongest of the windstorms, the 1962 Columbus Day blow, was the earliest (October 12), and this timing may not be an accident. The Columbus Day Storm, like several of the most powerful early fall events, began as a tropical storm that moved into the midlatitudes, changing its energy source from the warmth and moisture of the tropical ocean to the large temperature contrasts of the midlatitudes. In such cases it appears that some characteristics of the initial tropical disturbance are maintained and foster the development of the most intense storm circulations in more northern climes.

Pressure and Wind

For storms over the northeastern Pacific, the surface winds tend to move around the low-pressure center in a counterclockwise direction, with the

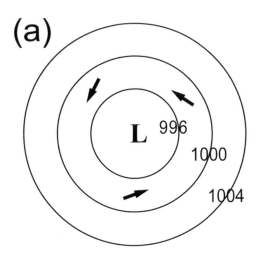

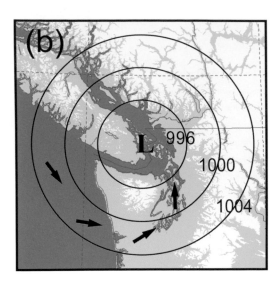

5.23. (a) For a storm over the ocean, winds roughly parallel the isobars (lines of constant pressure) and rotate around the low in a counterclockwise direction in the Northern Hemisphere. (b) Over land, particularly near terrain, winds tend to blow from high to low pressure. Illustrations by Beth Tully/Tully Graphics.

air moving nearly parallel to lines of constant sea-level pressure, known as isobars (figure 5.23). In contrast, over the mountainous land areas of the Northwest the air tends to flow toward lower pressure. What is the origin of these differences between land and water?

Wind is produced by differences in pressure, but this relationship has its complications. If we were not on a rotating planet, things would be easy. Air would generally blow from high to low pressure and the speed of the wind would grow as horizontal differences in pressure increased (figure 5.24a). However, we *do* live on a rotating planet and that changes things quite a bit. On such a planet there is an *apparent* force produced by rotation, called the *Coriolis force*, that acts upon

moving objects, such as air, so that it pushes an object to the right of its motion in the Northern Hemisphere (and to the left in the southern). Furthermore, the strength of the Coriolis force increases with wind speed and varies by latitude, being zero at the equator and reaching a maximum at the poles. The existence of such a force became obvious during the nineteenth century when long-range projectiles fired during European conflicts deviated to the right of their intended targets. The result of having a Coriolis force is that, away from obstructions, such as mountains, the winds blow nearly parallel to lines of constant pressure (as shown in figure 5.24b). Why is this the case?

Imagine we are on a planet that is not rotating, one in which the air is moving to the east from high to low pressure (again, see figure 5.24a). Then we slowly start rotating the planet. As a result of this rotation, a Coriolis force will begin to act on the moving air, causing it to deviate to the right. The air will continue to increase in speed as long as it has some motion from high to low pressure and the Coriolis force will increase with it (since the Coriolis force depends on wind speed). As the Coriolis force increases, the change in the air's direction will also increase until the air is moving to the south, since in that direction a balance can develop between the pressure force (from high to low pressure) directed to the left of the motion and the Coriolis force to the right of the motion (see figure 5.24b). Also in this direction, the speed of the air will not increase further, since the winds are blowing along a line of constant pressure. This balance between the Coriolis and pressure forces, known as *geostrophic balance,* is often apparent over the oceans. Over land and near mountains, as in much of the Pacific Northwest, the Coriolis force is less effective, since the air tends to be deflected by the terrain before there is sufficient time for the Coriolis force to work; thus, near mountains the winds tend to blow directly from high to low pressure. Such behavior is obvious in mountains gaps such as the Columbia Gorge or the Strait of Juan de Fuca, where the air moves up or down the gap toward the area of lowest pressure (chapter 7 discusses gap winds).

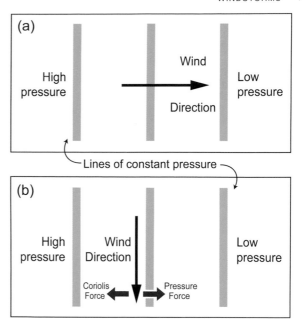

5.24. (a) On a non-rotating planet or near mountains, the winds tend to blow directly from high to low pressure. (b) In contrast, on a rotating planet and away from the equator in the Northern Hemisphere, the winds blow parallel to the isobars (lines of constant pressure), with high pressure to the right of the wind direction. Illustrations by Beth Tully/ Tully Graphics.

Windstorms and El Niño and La Niña

As mentioned in chapter 4, research has revealed a connection between the surface temperatures of the tropical Pacific and weather over the western United States. When tropical sea-surface temperatures are warmer than normal (El Niño periods), the Northwest tends to be warmer and slightly drier than usual during the winter, while cooler tropical sea-surface temperatures (associated with La Niña conditions) bring wetter than normal winters, with somewhat cooler and snowier conditions during the latter half of the winter. The tendency to alternate between El Niño and La Niña conditions over a period of three to seven years is termed the El Niño Southern Oscillation (ENSO), with the intervening "neutral" years having near-normal sea-surface temperature conditions (see chapter 11 for an elaboration of the El Niño/La Niña phenomenon).

ENSO appears to influence the frequency of major Northwest windstorms, with the greatest number of windstorms occurring in neutral (neither El Niño nor La Niña) years. To illustrate this, figure 5.25 shows the variation of sea-surface temperature for November through February for a region in the central tropical Pacific (called the Niño 3.4 area). Temperatures are shown as differences from the average (*anomalies* is the fancy term used by

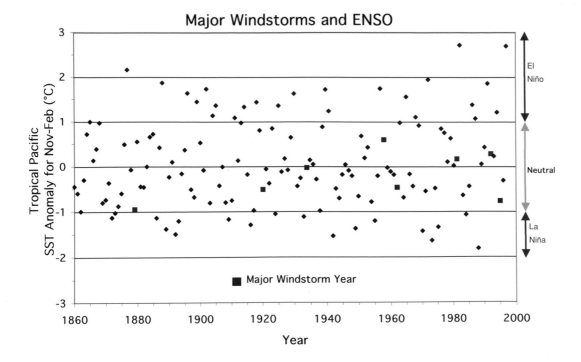

Major Windstorms and ENSO

■ Major Windstorm Year

5.25. Sea-surface temperature differences from normal (anomalies) over the eastern tropical Pacific (Niño 3.4 region) for November through February. Years with major windstorms are indicated by red squares, with non-windstorm years shown by blue diamonds. Major windstorms tend not to occur in El Niño and La Niña years, but prefer neutral years.

meteorologists) over the period 1856–2000. El Niño years are associated with temperatures more than 1 °C (2 °F) warmer than average, while La Niña temperatures are more than 1 °C (2 °F) cooler than normal. The sea-surface temperatures each year are indicated by blue diamonds except for the temperatures of major windstorm years, which are shown by red squares.[8] Clearly, the major wind-

storms were all in neutral years. Since there were only a small number of major windstorms during the past century, our sample is relatively small; thus, this ENSO/windstorm relationship should be considered suggestive, but not definitive.

The lack of major windstorms in El Niño years is not surprising; a number of researchers have found that El Niño periods are often associated with a "split flow" over the eastern Pacific, with the Pacific jet stream dividing and one portion going into Alaska and the other heading toward California. Such conditions tend to tear weather systems apart as they approach the Northwest—which lessens the chance for strong windstorm development. La Niña years often have periods with enhanced high pressure over the eastern Pacific (termed "ridging") that strengthens cool flow from the north. This situation is quite different than the typical windstorm pattern in which strong southwesterly flow heads directly into the Pacific Northwest. Since the transition between El Niño to La Niña occurs relatively slowly and can

---

8   The major windstorms that were selected all produced massive damage over the Northwest with winds of 60 mph and greater. The storms were January 9, 1880, January 29, 1921, October 21, 1934, November 3–4, 1958, October 12, 1962, November 13–15, 1981, January 20, 1993, and December 12, 1995.

often be predicted at least three to six months in advance, the connection between ENSO and major windstorms provides a several month "heads up" regarding the probability of a significant wind event.

Windstorm Fireworks

When a Northwest windstorm strikes at night, local residents are often provided with a colorful display of sky-filling light flashes accompanied by loud booms—reminiscent of summer fireworks. The cause of the bright lights and noise is the explosive opening of high-voltage fuses or "cut-offs" on local power lines (figure 5.26). As winds increase to about 30-40 mph, branches are blown against power lines or break and fall on the lines, causing them to be grounded or shorted. If the tree limb does not either burn through or fall away in a few seconds, the large current associated with the grounded circuit causes a nearby power line fuse to blow, as an explosive charge opens the circuit with a bright flash and a thunderous sound. The author has sat spellbound on several windstorm nights, watching the sudden illumination of low clouds by the colorful lights of the exploding fuses.

5.26. A power-line fuse or "cut-off." When a power line is grounded, a small charge on a fuse is ignited, resulting in an open circuit and a power outage for nearby homes.

With each flash a neighborhood goes dark, adding greatly to the entertainment. During the late nineteenth and early twentieth century, Northwest windstorms were often accompanied by a different type of light show—real fires. Since gas and wood-fueled fires were used extensively for heating and illumination, strong winds frequently resulted in structural damage that in turn initiated building fires.

6

SEA BREEZES, LAND BREEZES, AND SLOPE WINDS

ON MANY SUMMER EVENINGS, RESIDENTS OF THE EASTERN SLOPES OF THE

Cascades feel a welcome northwesterly breeze that descends from the mountains, bringing

some relief from the high temperatures of the day. In the Strait of Juan de Fuca, summer sail-

ors must be prepared for daily periods of strong winds from the west, which often are power-

ful enough for the posting of small-craft warnings. Along the Oregon coast, strong northerly

winds sandblast beachgoers during summer afternoons, but weaken at night. Over Puget

Sound, gusty northerly winds, known as the Sound Breeze, often push southward during

warm afternoons, making northward biking difficult but creating wonderful conditions for an

early evening sail. In the Columbia Gorge, westerly winds develop most summer afternoons, making the region a mecca for wind surfers. During a hike in the Blue Mountains a warm current of ascending air greets hikers on a rocky outcrop, while at night a cool breeze descends from the same hill, rustling the leaves of nearby trees.

Such local weather features in which winds vary during the day in a regular way are called diurnal (daily) wind circulations and are driven by differences in temperature between water and land or between mountain slopes and the nearby atmosphere. Commonly known as sea breezes, land breezes, or slope winds, these local circulations are controlled by the daily cycle of warming and cooling and are generally strongest during the summer, when heating by the sun is most intense. Diurnal winds play an important role in Northwest weather, and this chapter explains why they occur and some of their most dramatic local manifestations.

SEA AND LAND BREEZES

Sea and land breezes are forced by the differing temperatures of adjacent areas of land and water. During sunny days, particularly during the summer, the land heats up more than the water. The reason for this difference is that water has a large heat capacity, meaning that it takes a great deal of heat or energy to raise its temperature significantly. In addition, the sun's rays can penetrate a distance into water, spreading the heating over some depth. In contrast, the land surface has a relatively small heat capacity and the solar heating is absorbed in a shallow layer near the surface. Thus, with the same energy coming from the sun, the land surface warms substantially on a sunny day, while the change in surface water temperature is slight.

As the land warms, it heats the adjacent atmo-

sphere, causing the air to expand, its density to decrease, and the atmospheric pressure at the surface to fall.[1] With lower pressure over land and higher pressure over water, low-level air over the water moves toward the land, thus forming the sea breeze (figure 6.1a). The sea-breeze circulation extends above the surface, with rising air over land, seaward winds aloft, and sinking air over the water.

At night the opposite situation can occur. The heating from the sun has ended and both the land and water emit infrared radiation to space. As noted earlier, all substances (even people) emit radiation, with warm objects radiating more. If the atmosphere is relatively cloud-free, the radiation emitted from the surface will be lost to space, causing the surface to lose heat and cool. With its small heat capacity, the land cools much more rapidly than the water and thus near-surface air temperature drops more over land. Cooler temperatures result in a denser atmosphere and higher pressure over land than over water. Such a difference in pressure forces low-level air to move from land to water, producing a land breeze (figure 6.1b). Sea- and land-breeze circulations can occur anywhere

---

1   A more correct explanation is that the warming over land causes the atmosphere to expand, thus increasing the amount of air above a fixed altitude aloft. Since pressure is a measure of the weight of air aloft, pressure then increases aloft over land. As a result, at the higher elevation, pressure is larger over land than water, and this difference in pressure aloft causes air to move seaward at that level. As air is moved over the water aloft, pressure will tend to increase at low levels over the water, which in turn helps to drive the sea breeze toward land near the surface. Sea breezes aren't as simple as one might think!

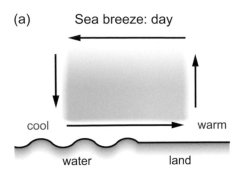

(a) Sea breeze: day

cool   warm

water   land

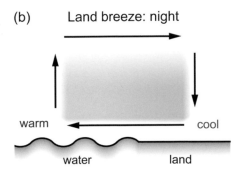

(b) Land breeze: night

warm   cool

water   land

6.1. Schematics of sea and land breezes. (a) Sea breezes occur when the land is warmer than the water, usually during the day, (b) while land breezes develop if the land becomes cooler than nearby water surfaces, generally at night. In the Northwest, land breezes are usually quite weak and sea breezes are mainly a summer phenomenon. Illustrations by Beth Tully/Tully Graphics.

that land and water are next to each other: along an ocean coastline, across the shorelines of large inland water bodies such as Puget Sound or the Strait of Juan de Fuca, or even near large lakes and rivers.

During the summer months, daytime air temperatures over land can warm into the 60s to 90s °F over western Washington and Oregon, while the surface temperatures of the Pacific Ocean, Puget Sound, and other major water bodies remain quite cool—generally below 55 °F. With such temperature differences between land and water, daytime sea-breeze circulations can develop. In contrast, summer land breezes are weak over the Northwest, since nighttime temperatures over land generally do not fall below the water temperatures. Land breezes are evident near shorelines during the winter periods of cold temperatures and relatively light winds when land air temperatures are 15 °F or more cooler than water temperatures.

SLOPE WINDS

A related type of local diurnal circulation is the slope wind. Consider a mountain slope with an adjacent low-elevation area (figure 6.2). On the mountain slopes, the sun heats the earth's surface, which in turn warms the nearby atmosphere; at the same elevation over the lowlands the atmosphere is not heated, since the atmosphere is relatively transparent to the sun's rays and the heating of the air is limited to the layer near the surface. This variation in the amount of heating at the same level produces a wind flow, related to the sea breeze, in which warm, less dense air moves upward along the heated mountain slope. After reaching the crest of the slope, the flow reverses direction, traveling back toward the lowlands. Finally, the air sinks over the lower elevations to complete the circulation. At night, the reverse situation (known as a drainage flow) occurs, with cold, heavy air sinking along the slopes and upward motion over the lowlands.

Most of us have experienced such upslope and downslope winds, particularly in the mountainous Northwest. On hikes in the mountains during a warm day one often feels the wind moving upslope

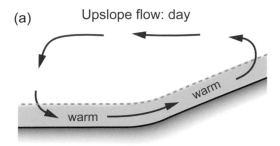

(a) Upslope flow: day

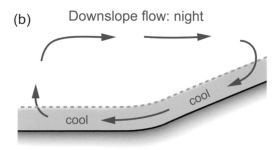

(b) Downslope flow: night

6.2. Schematic of upslope and downslope winds. Air in the red area is warmed by contact with the heated surface during the day, while air in the blue area is chilled by contact with the cooled surface at night. During the day, low-level air tends to move up the heated slopes (a), while at night cool air descends the slopes (b). Illustrations by Beth Tully/Tully Graphics.

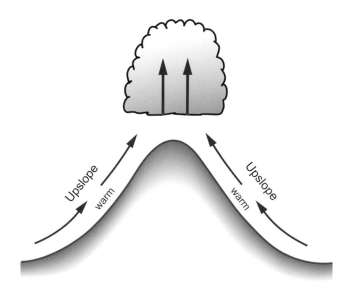

6.3. Upslope flows on both sides of a mountain can cause strong upward motion, clouds, and even precipitation over the higher terrain. Such clouds tend to form during the day when atmospheric conditions are favorable. Illustration by Beth Tully/Tully Graphics.

towards higher terrain. On the other hand, at night a cool breeze flows down local slopes. Such slope winds are frequently observed within the river valleys of the Cascades and Rockies during the warm season, with air moving upvalley during the day and downvalley at night. Downslope winds can provide natural summer air conditioning for those living at the base of slopes. The author's home is at the bottom of a hill on the northeast side of Seattle. Strolling at night with a thermometer in hand, I find that temperatures are often 3–6 °F warmer a few hundred feet up the block and perhaps 30–50

feet higher in elevation. Unfortunately, during the winter, the cool downslope winds have less welcome effects: not only do they bring greater heating costs, but smoke from fireplaces and wood-burning stoves tends to descend the slopes and collect in low spots, increasing the chances for asthma and other respiratory ailments. So buy or rent a home or apartment near the top of a hill if you are sensitive to poor air quality.

Upslope and downslope winds are sometimes revealed by clouds. For example, during late spring and summer days when upslope flow is pronounced, convective clouds can form over mountain crests as upslope flows on both sides of the crest converge near mountain peaks to produce strong upward motion (figure 6.3). Such strong

rising air can produce convective clouds, such as cumulus or cumulonimbus, when the atmosphere is nearly unstable so that a slight upward displacement will cause the air to become buoyant and rise on its own. Upslope clouds often build over the crests of the Cascades and Rockies and frequently produce showers at high elevations during the afternoon (figure 6.4). Sometimes such convection is extraordinarily impressive, with mountaintop cumulus or cumulonimbus towers stretching to the horizon.

THE DIURNAL WINDS OF WESTERN WASHINGTON

To illustrate the region's summer diurnal winds, figure 6.5 shows average surface winds[2] over western Washington for several times during a typical July day. At 6:00 AM there is a weak land breeze

---

2   In meteorology, surface winds are generally measured at 10 meters, roughly 30 feet, above ground level.

6.4. Growing cumulus clouds over the Cascades east of the Skagit Valley. Convective clouds form over the crest of the Washington Cascades and other Northwest mountains as daytime upslope flows on both sides converge near the mountain summits. Photo courtesy of Art Rangno.

along the coast, and modest winds flow eastward through the Strait of Juan de Fuca. By noon, the effects of surface heating over land are quite apparent. Along the coast, a sea breeze has developed as the winds strengthen and turn onshore. Winds over Puget Sound have also increased and are now directed from water to land. Such winds probably represent the combination of a sea breeze off Puget Sound and upslope flow climbing the Olympics and Cascades. By 3:00 PM (15 PST), near the time of warmest temperatures, the coastal sea breeze has strengthened further, with winds of 15–20 miles per hour at Hoquiam. The eastward flow in the Strait of Juan de Fuca has increased by at least 50 percent due to the contrast between heating and pressure falls over the Puget Sound interior (mainly land) and the cool waters and higher pressure over the

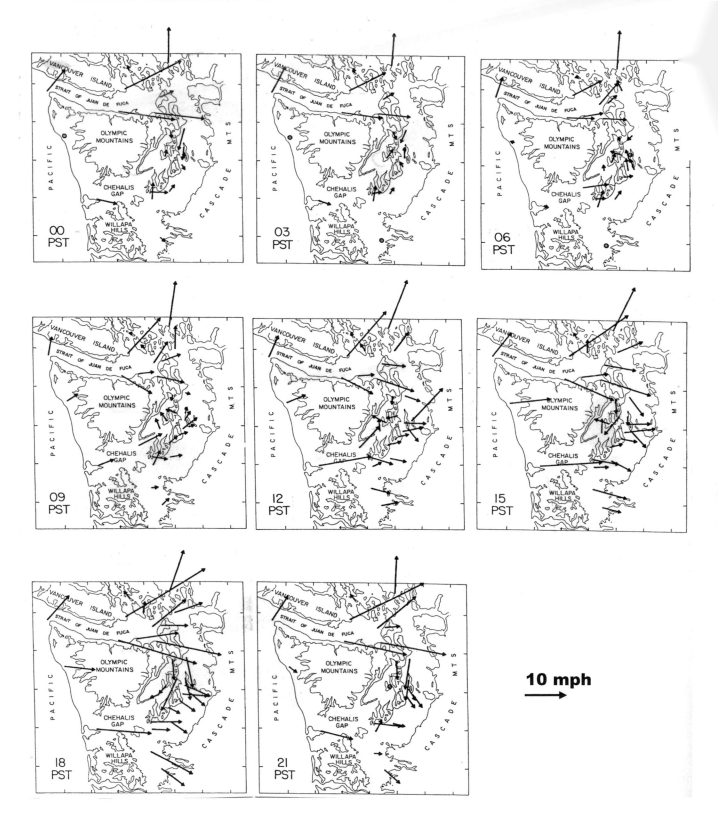

6.5. Average near-surface July winds over western Washington. The arrows point in the direction the wind is blowing and their length indicates the wind speed. Graphic from Mass (1982), courtesy of the American Meteorological Society.

Strait and Pacific Ocean. This air then heads southward into Puget Sound and is commonly known as the *Sound Breeze*. At 6:00 PM (18 PST), the winds in the Strait have increased to 20–25 miles per hour, with higher gusts, and the Sound Breeze intensifies and extends into central Puget Sound. By 9:00 PM (21 PST), the land has cooled considerably and both the coastal sea breeze and the Sound Breeze weaken. This weakening continues through midnight (00 PST), as does the slow decline of the westerlies (winds from the west) in the Strait of Juan de Fuca. These diurnal winds—the Sound Breeze, the Strait of Juan de Fuca late-afternoon westerlies, and the coastal sea breeze—have a large influence on life in western Washington.

The Sound Breeze brings cooler air—previously in contact with the water—into the Puget Sound region and helps keep the Puget Sound interior temperate during the summer. In contrast, western Oregon's land-locked Willamette Valley has no such natural air conditioning and is frequently 10–15 °F warmer than the Puget Sound lowlands. The Sound Breeze creates wonderful kite flying in Seattle parks during late summer afternoons and early evenings, until the winds collapse between 8:00 and 9:00 PM. After weather instruments were placed on a number of Washington State ferries, much was learned about the Sound Breeze. As illustrated in figure 6.6, the breeze is strongest over the Sound and the wind direction is northeasterly on the western side, northerly in the central sound, and northwesterly on the eastern side. It is believed that this splitting of the Sound Breeze is due to local sea-breeze winds, which move from water toward land. The total wind is a combination of winds of various origins and scales. Thus, over the eastern side of the Sound, the Sound breeze

from the north and the local eastward sea breeze along Puget Sound's eastern shoreline produce a total wind from the northwest.

The late afternoon/evening westerlies in the Strait of Juan de Fuca generally peak around sunset (8:00–9:00 PM during midsummer), with sustained winds on the American side of 15–20 miles per hour and gusts that can reach 30–40 miles per hour. In fact, these winds reach National Weather Service small-craft advisory levels (winds equal to or greater than 25 miles per hour) on most warm summer days and can be a safety issue for small boats. Strait summer westerlies are typically strongest in the central strait from Port Angeles to Victoria, are more intense on the Canadian side of the gap, and weaken rapidly over land. With strong westerly winds six to twelve hours per day, well-exposed vegetation near the Strait frequently has a windswept look, with a cant to the east (figure 6.7). Shallow sea fog often forms over the cool waters of the Pacific and is blown eastward by the Strait winds. This cool fog layer, coupled with strong winds, can be a major hindrance to navigation, as well as a serious visibility issue for the Whidbey Island Naval Air Station, located in the eastern terminus of the Strait. On the positive side, Strait fog provides needed moisture for vegetation within the rain-shadow region northeast of the Olympics.

Sea-breeze circulations have a profound influence on summer temperatures along the Northwest coast. The arrival of the sea breeze not only brings a substantial strengthening of the wind, but the onshore rush of air cooled by contact with the Pacific Ocean can "cap" the morning temperature rise, greatly limiting the daytime maximum temperature. To illustrate, figure 6.8 presents the temperatures and wind speeds along the central

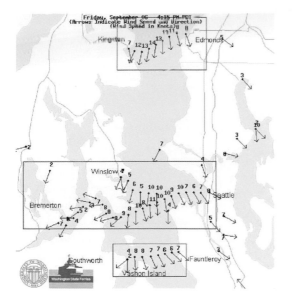

6.6. Winds observed on Washington State ferries and some nearby land stations at 4:15 PM on September 6, 2002. The arrows point in the direction to which the wind is blowing, and the numbers are wind speed in knots. One knot is approximately 1.15 miles per hour. A moderate Sound Breeze was occurring that afternoon. Note the change in direction of the Sound Breeze over central Puget Sound, ranging from northeasterly on the western side of the water to northwesterly on the Seattle side. Graphic from the Ferry Weather Web site (http://i90.atmos.washington.edu/ferry).

Washington coast at Hoquiam for a typical summer day. Winds are calm during the early morning hours, but strengthen during the morning as the land warms and a sea breeze develops. The sea breeze, which reaches a maximum in the afternoon and subsequently declines in the evening as the land cools, has a large effect on Hoquiam's temperature. Before the sea-breeze winds develop, temperature increases rapidly from approximately 45 °F to around 65 °F. However, as the sea breeze increases to near its maximum in the early afternoon, the temperature plateaus, with little change over the next four to five hours. It appears that the invasion of cool, marine air can overcome the additional heat received from the sun during the afternoon. In fact, an early afternoon maximum is characteristic of many Northwest coastal locations and differs greatly from the temperature evolution at inland sites where the highest summer temperatures usually occur between 4:00 and 5:00 PM.

6.7. Windblown trees along the Strait of Juan de Fuca near Sequim, Washington, show the effects of the strong westerly winds that are so prevalent during the summer months. The Dungeness Spit is in the background.

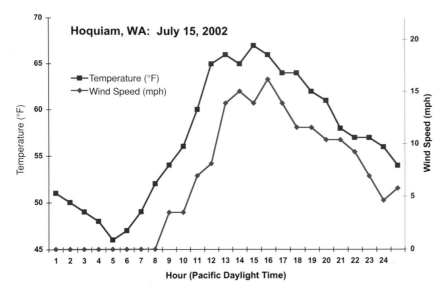

Hoquiam, WA: July 15, 2002

—■— Temperature (°F)
—◆— Wind Speed (mph)

6.8. Temperature and sustained winds (two-minute average, in mph) at Hoquiam, Washington (Boardman Airport), on July 15, 2002. The rise in temperature is abruptly shut down as the sea breeze brings cool marine air into the coastal zone.

6.9. Visible satellite photo at 1:00 PM on May 23, 2004. The leading edge of the sea-breeze winds (the sea-breeze front) is marked by a line of enhanced convective clouds inland from the coast. Arrows indicate the location of the sea-breeze front. Behind the sea-breeze front, the skies are nearly clear because the cool marine air prevents the formation of convective clouds. Image from the MODIS instrument on the NASA AQUA satellite.

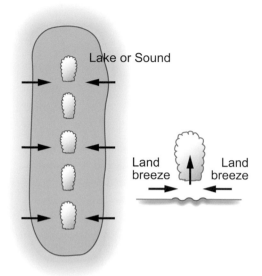

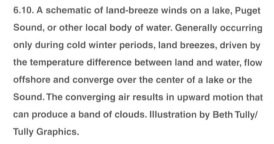

6.11. On November 28, 2006, a cold air mass had spread over the Puget Sound region in the wake of an Arctic Front from Canada. With very cold air over land and warmer temperatures over the relatively balmy Puget Sound, land breezes developed along the Sound shoreline and converged over the water, producing a north-south band of clouds. This picture was taken from the University of Washington that morning, with the band of clouds to the west over the Sound.

6.10. A schematic of land-breeze winds on a lake, Puget Sound, or other local body of water. Generally occurring only during cold winter periods, land breezes, driven by the temperature difference between land and water, flow offshore and converge over the center of a lake or the Sound. The converging air results in upward motion that can produce a band of clouds. Illustration by Beth Tully/ Tully Graphics.

The effects of local sea and land breezes are not limited to temperature and wind. If the air over land is nearly unstable, so that convection (cumulus and cumulonimbus clouds) can form if air is forced to rise, the abrupt lifting by the leading edge of the sea breeze, known as the *sea-breeze front*, can produce a line of clouds that moves inland with the ocean air (figure 6.9). Since convection requires a large temperature drop with height, the arrival of cool low-level air kills off the convective clouds behind the sea-breeze front, leaving a cloud-free zone in its wake. Clearing behind the sea-breeze front also occurs because of the sinking air over the water in a typical sea-breeze circulation (refer to figure 6.1).

In contrast, land breezes can produce a line of clouds over inland water bodies such as Puget Sound or Lake Washington during unusually cold winter nights when winds are relatively light. During such times, land temperatures can drop into the 20s °F or lower, while the water temperatures remain near 50 °F. Such temperature contrasts drive land breezes that develop on both shores and then converge over the central portions of a lake or sound (figure 6.10). As air converges, some of it is forced to rise, producing a band of clouds down the center of the body of water. The author has viewed such lines on many occasions over Lake Washington or the central Puget Sound, normally when cold air has settled in after a rare arctic outbreak (figure 6.11).

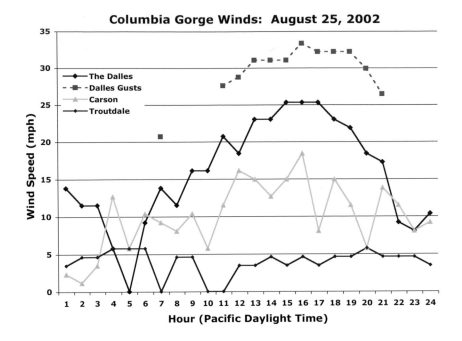

Columbia Gorge Winds: August 25, 2002

6.12. Sustained winds (two-minute average, in mph) on August 25, 2002, at Troutdale, Oregon; Carson, Washington; and The Dalles, Washington. Note the strong afternoon winds at Carson and The Dalles.

THE SUMMERTIME COLUMBIA RIVER GORGE "SEA BREEZE"

Although sea-breeze winds are produced by differing temperatures between land and water, a similar phenomenon can occur if two adjacent land areas are warmed at a different rate. This type of "sea" breeze explains the strong westerly winds that develop in the Columbia Gorge most summer afternoons—winds that have proved a boon for wind surfers.

During the summer, the region west of the Cascades is cooled by marine air from off the Pacific. In contrast, the land east of the Cascades is isolated from the marine air and is relatively arid and cloud-free. Thus, eastern Washington and Oregon heat up more than the western side of these states, resulting in enhanced pressure falls east of the mountains. This is true because air becomes less dense when it warms, and pressure at a point is a measure of the weight of the air above. Such

a warm area of lowered pressure is called a *heat trough* or *thermal low* and is most evident late in the afternoon, when temperatures are highest.

The Columbia River gorge is the only near sea-level passage across the Cascades. When pressure falls more to the east than to the west during a warm eastern Washington afternoon, a pressure difference builds across the Cascades. In response, air from the west side pushes into the only low-level conduit across the mountains (the Gorge) and accelerates as it moves eastward; thus, the strongest speeds are found in the central and eastern portions of this gap (see the discussion of gap winds in chapter 7). These westerly winds typically build in the afternoon, reaching sustained (averaged over two minutes) speeds of 10–20 miles per hour, with occasional gusts to 20 and 30 miles per hour. An example of typical summer gorge winds for a single day is shown in figure 6.12. While winds remain relatively weak (less than 6 miles per hour) at Troutdale on the western side of the Gorge, they

6.13. Wind surfing has become a major industry in the Columbia Gorge due to the dependable summer westerly winds that develop in the Gorge each afternoon. Photo courtesy of Alex Kerney.

progressively increase to the east at Carson (mid-Gorge) and The Dalles (at the eastern terminus of the Gorge). At The Dalles the sustained westerly winds accelerate to 25 miles per hour during the late afternoon, with gusts approaching 35 miles per hour. The winds rapidly decline during the evening and remain low overnight. These day-time westerlies have made the Columbia Gorge a world-class wind-surfing venue with moderate, dependable winds that move opposite to the westward river flow toward the Pacific. With such opposing effects, wind surfers can stay in the same general area without drifting downstream (figure 6.13).

THE DIURNAL NORTHERLY WINDS OF THE SOUTHERN NORTHWEST COAST

Summer coastal winds dramatically strengthen and turn more northerly as one proceeds south-ward down the Northwest coast, with maximum speeds over coastal southern Oregon and north-ern California. On many summer afternoons the coastal winds accelerate to 20 to 30 miles per hour, with higher gusts, making a walk on the beach more of an unpleasant sandblasting than a relax-ing stroll. Since the summertime winds in this area are generally from the north, coastal vegetation is frequently stunted in size and leans toward the south (figure 6.14).

The summertime (July–August) sustained (averaged over two minutes) wind speeds at four Northwest coastal locations are shown in fig-ure 6.15. All locations show large diurnal wind variations, with a rapid strengthening during the morning, maximum speeds during mid- to late afternoon, a rapid decline in the evening, and relatively light winds at night. Afternoon winds are far lighter at Hoquiam, on the central Washington coast, and Newport, on the north-central Oregon coast, than along the southern Oregon coast at North Bend and Gold Beach, where the sustained summer winds regularly peak near 20 miles per hour. Although the wind maxima at North Bend and Gold Beach are similar, the more southern location (Gold Beach) has far stronger winds at

6.14. A tree along the southern Oregon coast near Gold Beach is bent to the south due to the strong summer northerly winds of this area.

night. On a dozen or so days each summer, southern Oregon coast winds are much stronger than shown in the figure, with sustained northerly winds of 25–35 miles per hour and gusts between 35 and 45 miles per hour. Alternatively, some days are accompanied by coastal winds that are light or from the south.

Why are the summer afternoon winds strongest along the southern Oregon coast and generally from the north? The typical summer sea-level pressure pattern (figure 6.16) has a large area of high pressure over the eastern Pacific, while over land the pressure is lower. Low pressure is often found over the southwest United States and frequently extends northwestward to the California-Oregon border. This low pressure is the result of the intensely heated land of the Central Valley of California and Desert Southwest as well as compressional heating that occurs as air sinks down the mountain barriers of the western United States. With high pressure over the ocean and

lower pressure over the southwest United States, a region of large pressure contrast and associated strong winds is found along the northern California and southern Oregon coasts. A large east-west pressure change, with higher pressure offshore, is associated with winds from the north over the coastal and offshore waters (chapter 5 discusses the relation between pressure and winds). The high, blocking coastal terrain and the substantial north-south pressure variations of this area also contribute to the strong northerlies.

The northerly flow along the coast varies considerably during the day, with the sandblasting most active in the afternoon. The wind variation is driven by changes in the temperature and pressure over land. Warming temperatures result in falling pressure, and thus pressure differences between land and water increase during the day, which in turn leads to stronger winds.

The strong northerly winds along the southern

6.15. Sustained July–August wind speeds (two-minute average, in mph) along the Northwest coast.

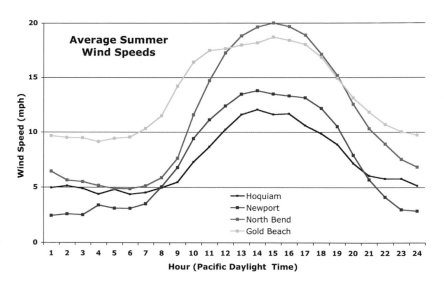

6.16. Sea-level pressure (black lines), temperature (shading from warm in red to cool in blue/green), and near-surface winds (wind barbs pointing in the direction the wind is going) for a typical summer day. Increasing winds are indicated by more lines on the tails of the wind barbs. High pressure over the eastern Pacific and lower pressure over the southwestern United States are the two major features. Note the large change of pressure and strong winds near the Oregon-California border.

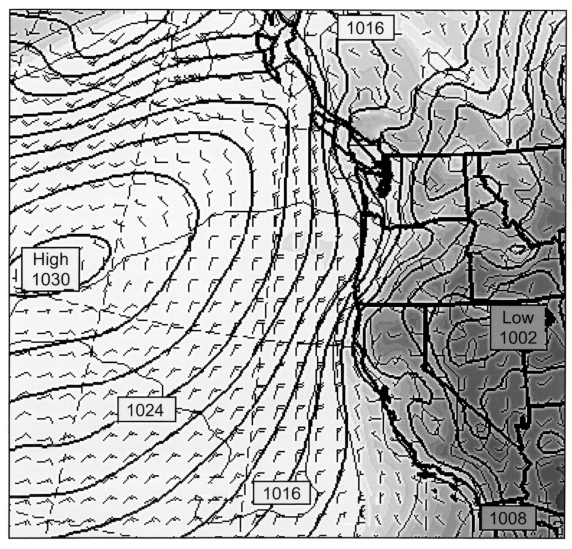

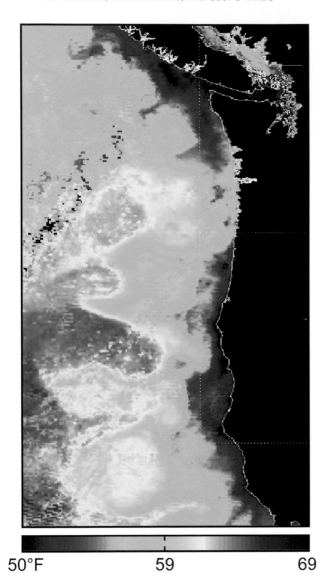

50°F   59   69

6.17. Sea-surface temperature analysis (°F) for September 17–24, 2006. Strong summer northerly flow produced coastal upwelling and cold water (blue colors) from California to southern British Columbia. Temperatures are far warmer offshore. Image from the MODIS instrument on the NASA AQUA satellite.

Northwest coast during summer have a large influence on the coastal ocean, causing cold water to rise to the surface (known as coastal upwelling). The upwelling, centered on a thin coastal strip

(approximately 20–100 miles wide) from northern California into British Columbia, plays a crucial role in Northwest fisheries because it brings up mineral-rich subsurface water. These minerals are needed by phytoplankton, the base of the oceanic food chain, which in turn support the fish population. An illustration of coastal upwelling is found in figure 6.17, which shows sea-surface temperatures during September 2006 that were observed from a NASA satellite. During that period, a narrow zone along the coast, centered on the southern Oregon and northern California shorelines, is roughly 10–15 °F colder than the water a few hundred miles offshore.

Why does upwelling occur? As noted above, the Northwest coastal zone, and particularly northern California and southern Oregon, experience strong northerly winds during the summer because of high pressure offshore and low pressure over the heated land. As these strong northerly winds flow over the Pacific Ocean, they tend to push the water to the south (the same way one can make water move by blowing on it). As the surface water starts to move toward the equator, the effect of the earth's rotation (the Coriolis force) pushes the water to the west (the Coriolis force tends to deflect to the right of the current's direction of motion). As shown in figure 6.18, with surface water moving away from the coast, and land to the east, water must move upward (upwelling) in the coastal waters. With water temperature decreasing with depth, such upwelling creates a narrow coastal zone of cold surface water. Since northerly flow is mainly limited to the summer, the cold coastal water is a warm-season phenomenon, one that is good for fish but unpleasant for swimmers and surfers.

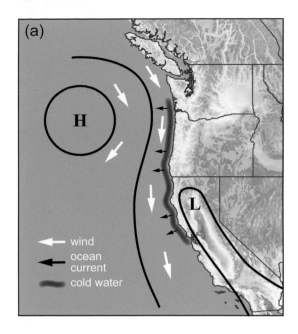

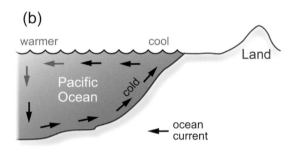

6.18. (a) Northerly winds along the Northwest coast result in offshore flow in the coastal ocean. (b) This offshore flow causes cold water from below to be drawn to the surface in the coastal zone. Illustrations by Beth Tully/Tully Graphics.

SLOPE WINDS AND THE NORTHWESTERLY WIND SURGE OF EASTERN WASHINGTON

Much of eastern Washington can be described as a topographic bowl. At the bottom of the depression, the Tri-Cities (Richland, Pasco, and Kennewick) are only a few hundred feet above sea level, while the surrounding terrain (the Cascades, the Okanogan Highlands, the

Rockies, and the Blue Mountains) rises to thousands of feet. Considering this basin topography, one would expect summertime upslope winds on the heated slopes during the day and downslope or drainage winds of cool air at night. Such slope flows do exist, but with an interesting twist.

Figure 6.19 shows the monthly average surface winds over eastern Washington for July 1950. At 12:00 and 6:00 AM, there are downslope winds produced by cool air draining toward lower elevations. In contrast, the noontime winds indicate a complete wind reversal, with air heading toward higher terrain. At 6:00 PM, when temperatures are still quite warm and upslope flow should be continuing, strong northwesterly winds push across the Cascades and descend the eastern slopes of the barrier into Ellensburg, Yakima, and the Tri-Cities. Such strong northwesterlies progressively push eastward between 4:00 and 8:00 PM on many summer days and last for approximately six hours. At the surface, the late-day northwesterlies often reach sustained speeds of 15–25 miles per hour, with occasional gusts to 30–40 miles per hour over the eastern Cascade foothills (Ellensburg and Wenatchee). The strength of the northwesterlies weakens as they reach the central portion of the basin near the Tri-Cities.

What is the origin of these northwesterlies? Careful examination of the development of these winds, coupled with numerical simulation experiments, reveal that they represent a regional "sea breeze" between western and eastern Washington. As noted in the discussion of the Gorge winds, the marine influence keeps summer temperatures moderate west of the Cascades, while eastern Washington, isolated from the Pacific or other

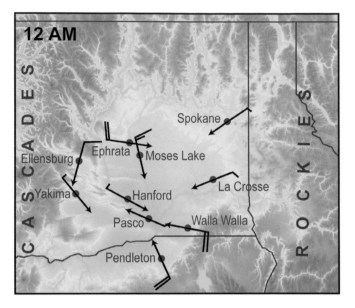

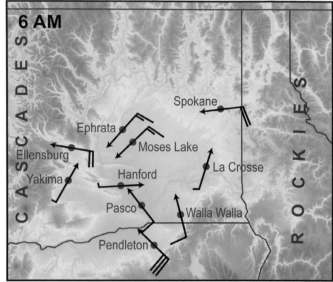

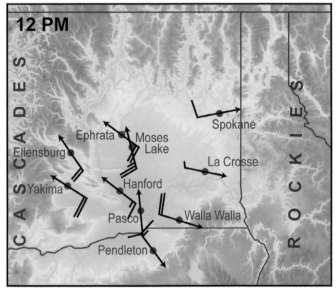

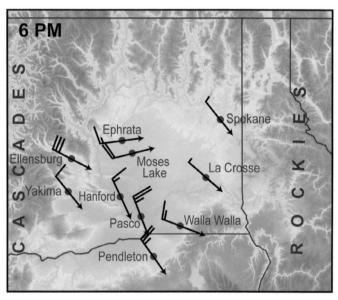

6.19. Average surface winds over eastern Washington for July 1950. The wind barbs point in the direction to which the wind is blowing. On the tails, long lines indicate speeds of 2.25 miles per hour and short lines half that speed. Two long lines indicate 4.5 mph, etc. Winds tend to be upslope during the day and downslope at night, except for the strong northwesterlies that push in from the west during the evening. Data from Staley (1959). Illustrations by Beth Tully/Tully Graphics.

significant water bodies, is much warmer. As a result, a difference in pressure can develop across the mountains, with lower pressure over the heated interior. Just as the traditional sea breeze is driven by horizontal temperature and pressure differences, a regional sea breeze develops with air flowing from the cool region (western Washington) toward the warmth and lower pressure of eastern

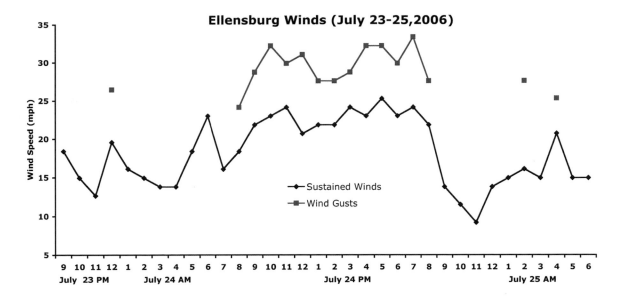

6.20. Sustained winds (two-minute average, in mph) and gusts at the Ellensburg Airport during July 23–25, 2006. Note the strong acceleration during the morning as pressure fell over eastern Washington due to surface heating.

Washington. Such cross-Cascade flow strengthens during the late afternoon when daytime temperatures reach their peak over eastern Washington and associated cross-Cascade pressure differences are largest. This cross-mountain sea breeze is further enhanced when the marine air is particularly deep west of the Cascades, such as after a marine or onshore push.

A recent example of strong summertime northwesterly flow occurred on July 23–25, 2006. This period was an unusually warm one, with temperatures in the Columbia Basin reaching over 100 °F, while 80s and 90s °F were observed west of the Cascades. As the basin warmed during the morning, pressure decreased, resulting in an increasing pressure difference and strengthening winds across the mountains and along the eastern slopes. At Ellensburg, winds of approximately 15 miles per hour during the early morning of July 24 were followed by a substantial speed-up during the late morning and afternoon to sustained winds of 20–25 miles per hour, gusting to 25–35 miles per hour (figure 6.20). These strong northwesterlies

began earlier than normal due to the extreme temperatures. If such winds had occurred over water, the National Weather Service would have issued a small-craft advisory. Later that evening, as air temperatures cooled over eastern Washington, the winds declined substantially.

In contrast to its value for wind energy generation, the strong northwesterlies have a more ominous side during the summer wildfire season. Such strong flow can often stoke minor fires, initiated by lightning or human inattention, into exploding and fast-running conflagrations and can make fighting established fires difficult, if not impossible. Fortunately, the regular timing of the northwesterlies and the apparent skill of high-resolution weather-forecasting models in predicting their development and evolution promise to provide fire fighters with useful guidance.

7

COASTAL WEATHER FEATURES

||||||||||||||||||||||||||||||||||||||||||||||||||||||||||||||||||||||||||||||||||||||||||||||||||||||||||||||||||||||||||||||||||||||||||||||||||||||||||||||||||||||||||||||||||||||||||||

A FASCINATING COLLECTION OF LOCAL WEATHER FEATURES OCCURS OVER THE

coastal zone of the Pacific Northwest, particularly when mountains are nearby. In addition to

the regional sea breezes discussed in the previous chapter, there are many other important

weather phenomena along the Pacific coastline and over inland bodies of water, such as the

Strait of Juan de Fuca and Puget Sound. For example, there is the coastal "banana belt" over

the southern Oregon shore, where extraordinarily warm temperatures can occur throughout

the year when air moves westward toward the ocean. In contrast, the Northwest has a ready

air conditioner, commonly known as the onshore or marine push, which periodically inun-

dates the coastal region with cool marine air during the warm season. Also during late spring and summer, strong winds can push northward along the coast as alongshore surges bring dramatic changes in clouds and temperature. Strong winds are not limited to the immediate coast, but also occur in the Juan de Fuca and Georgia straits, where damaging gusts can reach 70 to 80 miles per hour. Finally, the Puget Sound region is often in cloud and precipitation due to another coastal feature, the Puget Sound Convergence Zone, which is established when Pacific air is deflected around the Olympic Mountains, only to converge over the Sound.

THE BANANA BELT OF THE SOUTHERN OREGON COAST

When one thinks of the Pacific Northwest coast, warmth is generally not the first thing that comes to mind, but in fact, the southern Oregon coast is often visited by relatively torrid weather—even during the middle of the winter. Brookings, on the coast just north of the California-Oregon border, holds the record for all-time warmth for *any* Oregon location for every month from November through March. For example, on February 27, 1985, Brookings reached 81 °F, the highest temperature ever recorded in Oregon during the month of February. In contrast, the remainder of the Oregon coast only climbed into the 50s and 60s °F that day, with even cooler temperatures over the interior of the state (figure 7.1a). Even in the depth of winter, temperatures in the 70s °F are not unusual along the southern Oregon coast. This warmth— often called the Brookings effect—has caused the coastal area near Brookings to be known as the banana belt of Oregon, with the local

chamber of commerce hawking the area's "tropical" environment.

The origin of the banana belt can be understood by referring to a topographic map (figure 7.1b). While most of the Northwest coastline has relatively low (2,000–3,000 feet) and narrow coastal mountains to the east, the southern Oregon coast is on the western flanks of the high (4,000–6,000 feet) and wide Siskiyou/Klamath Mountains. Occasionally, the typical onshore winds, blowing from the southwest to northwest, are replaced by offshore flow from the northeast to southeast. This situation usually occurs as high pressure builds over eastern Oregon at the same time that low pressure extends northward up the California coast into southwestern Oregon (figure 7.2). In such cases, air descends from high elevations over the Siskiyou/Klamath Mountains to near sea level over coastal southern Oregon. As the air subsides, it experiences higher pressure and thus is compressed and warmed. This process of downslope warming, described in the previous chapter, is analogous to the warming produced as a tire pump compresses air.

Although the southern Oregon coast provides the most dramatic example of downslope warming, similar effects occur less frequently at other locations across the Northwest. For example, when the winds aloft are from the northeast, extraordinarily high temperatures are often observed at Forks, Washington, which is then downwind of the Olympic Mountains. Downslope warming is also frequently evident north of the Blue Mountains of southeast Washington during strong southerly flow. The western foothills of the Cascades, such as the area around North Bend, Washington, are often

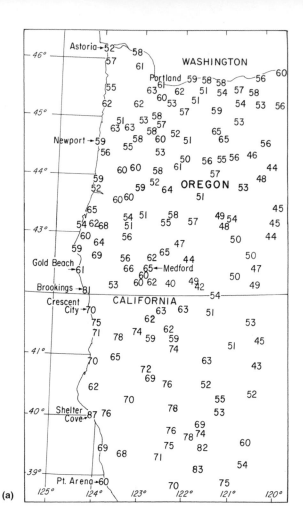

(a)

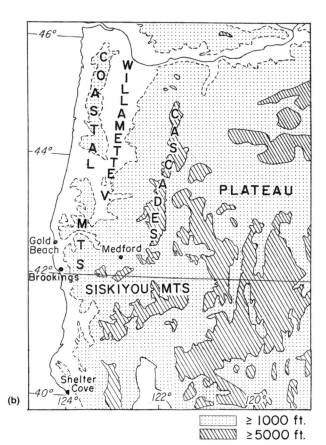

(b)

≥ 1000 ft.
≥ 5000 ft.

7.1. (a) Maximum temperatures (°F) on February 27, 1985, over western Oregon and northern California. Note that Brookings, Oregon, reached 81 °F, while Gold Beach, 20 miles to the north, only climbed to 61 °F. (b) Topographic map of Oregon and northern California. Brookings is to the west of the relatively high and wide Siskiyou Mountains. Graphics from Mass (1987), courtesy of the American Meteorological Society.

7.2. Sea-level pressure and surface observations at 4:00 PM PST on February 27, 1985. High pressure is centered over eastern Oregon, while a trough of low pressure extends northward along the California coast. At each location, temperature is shown (upper left) as well as the wind speed and direction, indicated by wind barbs, which point in the direction to which the wind is blowing. On the wind barbs, a long line on the tail indicates roughly 11 miles per hour and a short line 6 miles per hour. Graphic from Mass (1987), courtesy of the American Meteorological Society.

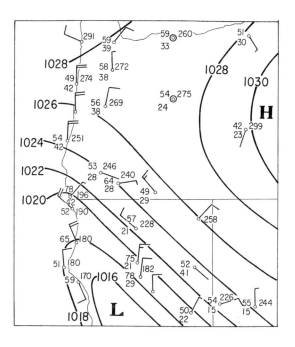

the warmest locations west of the Cascade crest when moderate to strong easterly flow is occurring over the region.

THE SUMMERTIME ONSHORE PUSH OF MARINE AIR

The most important Pacific Northwest weather phenomenon during the late spring and summer is the onshore surge of cool, cloudy, marine air that envelops the region west of the Cascade crest after a period of above-normal temperatures. Known as the *onshore* or *marine push*, this change from dry, warm offshore winds to a cooler, moist onshore flow is often accompanied by strong gusty winds, rapid temperature falls, large increases in humidity, and the influx of low clouds. Occasionally, a marine push is accompanied by lightning and thunder when warm, less stable air is pushed upward by the invading cool flow.

Marine pushes are most pronounced west of the Cascades and affect western Oregon, Washington, and southwestern British Columbia. In particularly strong events in which the invading marine air is deep, some of the ocean air can push across the Cascades, resulting in strong winds over the eastern slopes and a moderation of temperatures over eastern Washington and Oregon. Marine pushes are most frequent from May through September and most often make their entrance over Puget Sound and the Willamette Valley between 6:00 and 10:00 PM. One of the great pleasures of living in the Northwest is to experience a marine push after a hot day: the wind chimes sound, the leaves begin to rustle, the air quickly cools, and a night of comfortable sleeping is anticipated.

The typical weather evolution that produces a marine push is shown in figure 7.3. Figure 7.3a shows the normal summertime pattern over the Northwest: the East Pacific High is centered offshore, resulting in weak onshore flow of marine air and moderate temperatures over western Oregon and Washington. Low pressure associated with very warm temperatures is found over the interior of California. In the next step (figure 7.3b), high pressure builds over the interior of the continent as a portion of the East Pacific High moves eastward, resulting in offshore (westward) winds over the Cascades. Such winds move warm air from the continental interior toward the coast, and further warming occurs as the air descends the western slopes of the Cascades. As sinking warm air displaces cool marine air, western Oregon and Washington rapidly warm to temperatures far above normal. This warming causes pressure to fall west of the Cascades and along the coast, since warm air is less dense or heavy than cold air, resulting in the northwestward extension of the California low-pressure area (also known as the heat or thermal trough). A tongue of coastal stratus often follows the heat trough up the coast and can be seen on weather-satellite pictures (see figure 7.4). This stratus is associated with coastal air that is moving northward toward the lowest pressure on the coast; such air has a trajectory over the cool water, in contrast to the warm, dry, offshore flow found within the heat trough.

As long as the high pressure over the continental interior is strong enough to produce vigorous offshore flow, the warm temperatures in the heat trough west of the Cascades continue, as shown in figure 7.3c. The coastal stratus and associated cool marine air along the coast often deepen during this period, causing higher pressure to build along the

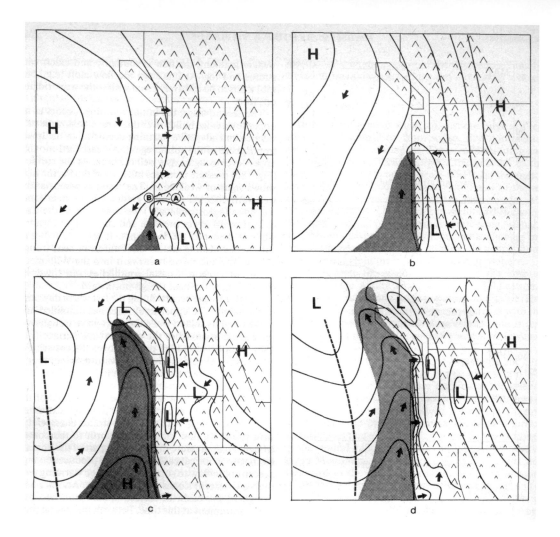

7.3. Schematic of the weather patterns associated with summertime marine pushes. Solid lines are lines of constant pressure, the arrows indicate low-level wind direction, and the shading denotes areas of low clouds. (a) Normal summer situation, with high pressure offshore and weak onshore flow. (b) Initial warm-up stage as high pressure building inland causes offshore and downslope flow, with the California heat trough moving northward. (c) Maximum temperature stage, with the heat trough extending into southern British Columbia. Low-level clouds and coastal high pressure have built northward along the coast as a Pacific disturbance approaches from the west. (d) Beginning of the marine push. The approach of the Pacific disturbance and the weakening or eastward movement of the interior high-pressure area allow marine air to surge into western Oregon and Washington. Graphic from Mass, Albright, and Brees (1986), courtesy of the American Meteorological Society.

coastal zone. This coastal high-pressure zone is enhanced if a weak Pacific disturbance approaches the coast, banking more cool air against the coastal mountains. Finally, the interior high-pressure area moves eastward, since most weather systems move east in the midlatitudes. As a result, the protective offshore flow over the coastal zone is lost, and the heat trough "jumps" over the Cascades into eastern Washington (figure 7.3d).[1] Marine air along the

---

1   With the high-pressure area and associated westward flow moving eastward, the sinking and warming are now over the western slopes of the Rockies. Thus, the greatest warming and pressure falls move eastward into eastern Washington.

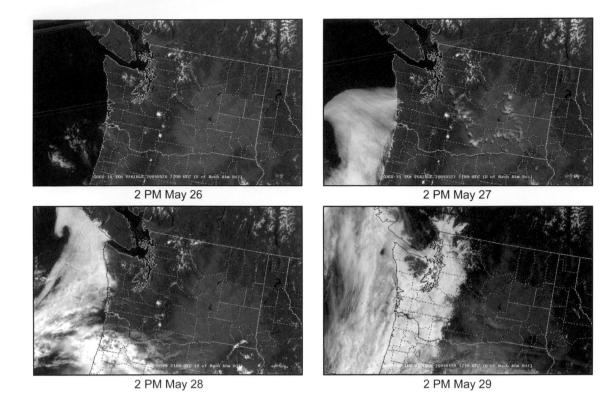

2 PM May 26

2 PM May 27

2 PM May 28

2 PM May 29

coast then surges inland toward the low pressure to the east, and the marine push is underway.

Weather-satellite pictures for May 26–29, 2005, illustrate the profound changes in cloud cover that accompany a strong marine push (figure 7.4). In the first image, the region is nearly cloud-free as warm, offshore flow dominates the entire area. The next day, stratus starts moving up the coast as the trough of low pressure extends northward into western Washington and southern British Columbia. The subsequent image immediately precedes the marine push: low clouds have spread along the entire coast and some appear to be moving inland on the central Washington coast. At the same time, higher clouds associated with an approaching disturbance extend inland over northwest Oregon and southwest Washington. Finally, the last picture shows the region after the push

7.4. Visible satellite imagery over Washington State and environs at 2:00 PM each day from May 26 through May 29, 2005, illustrating a typical marine-push event. Image from National Oceanic and Atmospheric Administration/National Weather Service GOES-West weather satellite.

has occurred, with all of western Washington and Oregon covered by low clouds.

The one-day drop in maximum temperature associated with marine pushes is sometimes as large as 20–25 °F. In fact, the day-to-day temperature falls associated with this summer onslaught of marine air are far larger than the temperature drops associated with strong Pacific fronts in winter. During winter, air generally travels eastward over the vast Pacific Ocean, whose relatively uniform and mild temperatures modify the lower atmosphere, reducing the temperature contrasts

7.5. Altocumulus castellanus over Seattle. During the summer, such midlevel instability clouds often indicate the approach of a Pacific weather disturbance that can initiate an onshore marine push.

across fronts originating over Asia and Alaska. Thus, most wintertime frontal passages only result in a cooling of a few degrees at most. In contrast, surface air temperatures in the offshore flow preceding a marine push can reach 80–90 °F (or even more), while the marine air following the push has been over the relatively cool waters of the eastern Pacific (roughly 50–55 °F); thus, the transition between these two air flows can result in extremely large temperature changes. The cool Pacific Ocean is the natural air conditioning system of the Pacific Northwest, and the marine push represents the cooling switch being flipped on after a period of warmer than normal weather.

There are a number of signs of an impending marine push, some evident in the sky and some indicated by the barometer. Most push events are associated with an approaching weather disturbance. Such weather systems cause upward motion that can produce altocumulus castellanus clouds—cumulus clouds that form several thousand feet above the surface in towerlike shapes (figure 7.5). Another sign of an impending push is the increase in the pressure difference between the coast and the western Washington and Oregon interiors. If the pressure on the coast gets substantially higher than over the interior (by 3–4 millibars, or 0.1 inch of mercury), a moderate or strong push is probably on the way. In addition, the arrival of the marine air is often signaled by a sudden decrease in visibility since the high humidity associated with the marine air causes tiny particles, called condensation nuclei, to swell as water vapor condenses on them. Marine air often has a large number of salt particles, produced when water droplets generated by breaking waves evaporate. Salt particles are very effective in absorbing water vapor, which is why salt shakers often clog when humidity is high. Thus, marine air, with both many salt particles and high humidity, is usually associated with numerous large particles that can obscure visibility. Finally, high-resolution computer forecasting models, which simulate atmospheric evolution, generally do an excellent job predicting marine push initiation and evolution and have become the key tool for predicting these events a day or two in advance.

ALONGSHORE SURGES OF THE NORTHWEST COAST

May 16, 1985, began as a beautiful day on the northern Oregon and Washington coasts, with clear skies and unusually warm temperatures. With northeasterly winds blowing warm air offshore, temperatures at coastal stations climbed into the mid- to upper 80s °F by midafternoon,

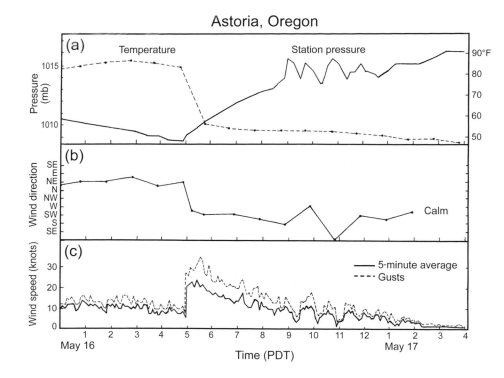

7.6. Plot of (a) temperature and pressure, (b) wind direction, and (c) wind speed at Astoria, Oregon, during May 16-17, 1985. Dramatic changes in temperature, pressure, and wind were evident around 5:00 PM PDT as the alongshore surge passed that location. Graphic from Mass, Albright, and Brees (1986), courtesy of the American Meteorological Society.

with Astoria, on the northern coast of Oregon, hitting 87 °F, an all-time record for the date (figure 7.6). At Astoria, the first sign of a major change was the appearance of a wall of stratus on the southern horizon around 5:00 PM. As the clouds approached, the winds shifted 180 degrees: from the northeast at roughly 12 miles per hour to the southwest, with gusts to 35 miles per hour. As the winds reversed, temperatures plummeted, dropping from 86 °F at 6:00 PM to 57 °F an hour later, and the sky turned completely overcast. While dramatic changes were occurring along the coast, temperatures remained warm in the nearby Willamette Valley and the interior of western Washington, with both regions enjoying the temporary protection of the low mountains that parallel the Pacific coast.

Such large and dramatic shifts in coastal weather from warm, offshore flow to cool, cloudy, and often windy weather from the south are known as *alongshore surges*. Strong events, like the May 1985 case, typically occur one to two times per year along the Northwest coast, with weaker events occurring perhaps once a month during the warm season. Although alongshore surges are most frequent and energetic during the late spring and early summer, they can strike in any month between March and October. On a weather-satellite image, alongshore surges are usually associated with a narrow coastal tongue of stratus, 50–150 miles wide, that moves northward along the coast (figure 7.7). The transition from offshore (easterly) flow north of the surge to the southerly flow of the surge coincides roughly with the northern boundary of the coastal clouds. The skies are generally clear in the offshore flow north of the surge, since the easterly flow has warm, dry continental origin. The large

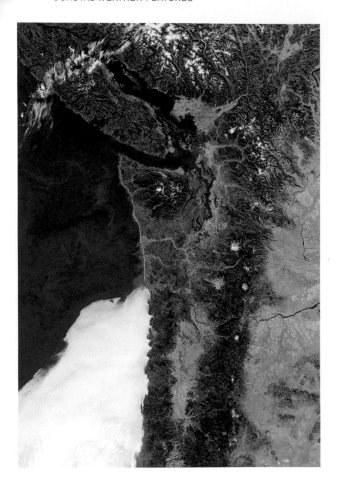

7.7. Satellite image of an alongshore surge moving northward along the Oregon and Washington coasts at 2:00 PM on September 27, 2003. Strong southerly flow is found within the low stratus clouds, while to the north, warm, offshore flow is causing clear skies. Image from the NASA MODIS AQUA satellite.

weather changes associated with an alongshore surge are relatively shallow, generally limited to the lower 3,000–5,000 feet of the atmosphere.

With a transition from warm, offshore flow to cool, marine air from the south, alongshore surges usually bring large and sudden declines in temperature. In the strongest surges, the cold, dense air from the south moves northward as an intense *gravity current*, which is similar to the rush of water

after a dam breaks. In such cases, the change in wind and temperature can occur in a matter of minutes. The Northwest coast is not the only location that experiences such intense coastal transitions; similar features also occur over southeast Australia (known as the *southerly buster*), along the coast of Chile, and on the coast of South Africa.

Why do coastal southerlies occur periodically along the Northwest coast in summer? The explanation has much in common with that of the marine push, since coastal surges often occur prior to the onshore movement of cool, marine air. When high pressure builds inland, offshore flow produces warming and pressure falls along the coastal zone. As a result, the California heat trough of low pressure moves northward up the Northwest coast (refer to figure 7.3). This trough produces a difference in pressure along a portion of the coast, with lower pressure in the trough and high pressure to the south. Near mountains, air tends to accelerate from high to low pressure, so that west of the coastal terrain the air moves northward in response to the pressure difference, producing the alongshore surge.

The rapid changes in wind and visibility associated with alongshore surges can greatly affect marine interests, particularly sailboats and other recreational boating, and can cause a dramatic *warming* of the coastal waters. Why warming? As described in the previous chapter, high pressure builds in the eastern Pacific during the summer and results in generally northerly winds along the Northwest coast. Such northerly winds cause upwelling over the coastal ocean, which forces cold water to rise to the surface. When the winds reverse to southerly, the upwelling is shut down, causing the sea surface to warm.

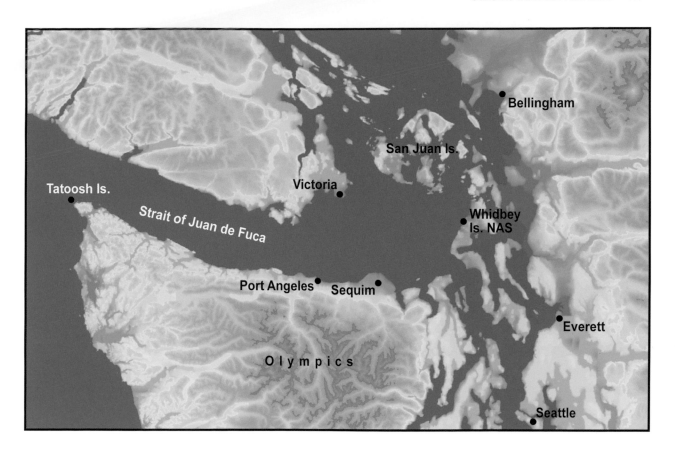

7.8. Terrain and major geographical features of the Strait of Juan de Fuca and environs.

THE WINDS OF THE STRAIT OF JUAN DE FUCA AND ENVIRONS

Strong winds are a common occurrence in and near the Strait of Juan de Fuca, a sea-level passage between the mountains of Vancouver Island and the higher Olympic range (figure 7.8). Summer days often bring westerly winds of 20–30 miles per hour, occasionally with gusts to 40–50 miles per hour, and in the winter gusts of 60–85 miles per hour have been observed. As explained below, the greatest wind speeds are generally *not* found in the relatively narrow central Strait but rather at its eastern or western flanks.

Gap Wind 101

The flow through the Strait of Juan de Fuca is an example of a gap wind, in which air is forced to move through a relatively narrow passage between two mountain barriers. Gap winds are often poorly explained, even in some introductory meteorological textbooks. Frequently, the idea of a venturi or funnel effect is cited, in which air is accelerated as it enters a constriction. The typical explanation is that a small opening with faster flow will allow as much air to move through as a larger area with slower-moving air. Thus, one would expect air to accelerate as it passes through the narrowest part of the gap. Using similar reasoning, the air would subsequently slow after passing the constriction. Although such venturi effects can be extremely

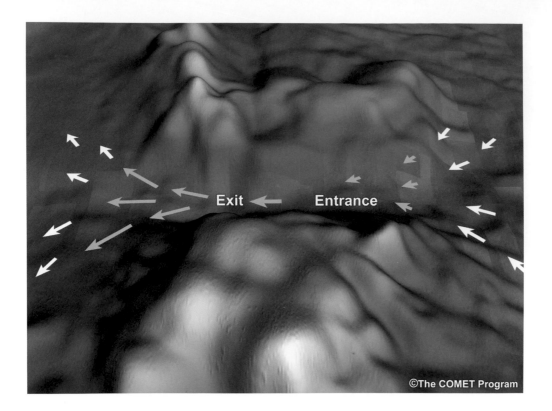

Exit ← ← Entrance

©The COMET Program

7.9. In the usual discussion of flow through mountain gaps, the venturi effect is often cited as the explanation, suggesting that the strongest winds should be found in the narrowest part of the gap. In contrast, larger-scale gaps such as the Strait of Juan de Fuca typically have their strongest winds at the exit of the gap. Graphic courtesy of the UCAR COMET program, with special thanks to Steve Deyo. The arrows indicate low-level winds with their length proportional to the speed.

important for smaller gaps or even for the spaces between buildings, they are generally of secondary importance for gaps the size of the Strait of Juan de Fuca and the Columbia Gorge. For such larger gaps, the strongest winds are typically found at the *exits* of the gaps, not at their narrowest points (figure 7.9).[2]

Why are winds strongest on the exits of large gaps and not at their narrow centers? First, there are often large-scale pressure differences across gaps, with high pressure on one side and low pressure on the other. For example, over the Northwest, a strong high-pressure area over the interior and an approaching Pacific storm (with associated low pressure) can create large pressure differences across regional gaps, such as the Strait of Juan de Fuca. Since air accelerates as it moves from high to low pressure,[3] the winds will tend to speed up across the entire gap. Second, there is often cold air on the high-pressure side of the gap. As air approaches the gap, the depth of the cold air generally increases on the windward side of the bar-

2   Air enters the gap through the gap *entrance* and leaves the gap at its *exit*.

3   This is analogous to a toothpaste tube. When one squeezes the tube, a difference in pressure occurs between inside the tube and outside, accelerating the toothpaste out the opening.

rier, due to the blocking effects of the surrounding terrain. This increase in the depth of cold, dense air produces higher pressure near and upwind of the center of the gap that slows the air as it enters the gap (figure 7.9). In the gap exit region, the rapid widening of the gap causes the gap flow to spread horizontally and rapidly thin. This thinning of the dense, cold air results in rapidly falling pressure in the gap exit, since pressure is simply the weight of the air above. As a consequence of the large pressure difference created by the thinning cold air, winds accelerate over the exit region.

Easterly Wintertime Gales in the Strait of Juan de Fuca

An early study of the winds in the Strait of Juan de Fuca found that when air blows from the east, strong winds are a common occurrence over the western end of the Strait near Tatoosh Island (see Reed 1931). In fact, over a five-year period there were more than two hundred easterly wind events that exceeded 40 miles per hour, with the vast majority during the winter. The combination of such powerful winds, a protruding rocky coast, and frequent clouds and fog makes the area around Tatoosh Island highly dangerous to the mariner. To illustrate the risks, during the days of sail (1830–1925) there were 137 major marine losses near the entrance to the Strait of Juan de Fuca, a region that became known as the graveyard of the Pacific.

Modern technology provides graphic illustration of the easterly winds in the Strait. Specifically, a new generation of satellites carries highly sensitive radars that can measure the size of waves on the water surface. Making use of the relationship between wind speed and the amplitude of small surface waves, a wind-speed map during an easterly wind event through the Strait was created (figure 7.10). Note that although the wind speeds do increase in the narrow part of the Strait, the strongest winds, indicated by red and bright yellow colors in the figure, are found over the far western end near Tatoosh Island.

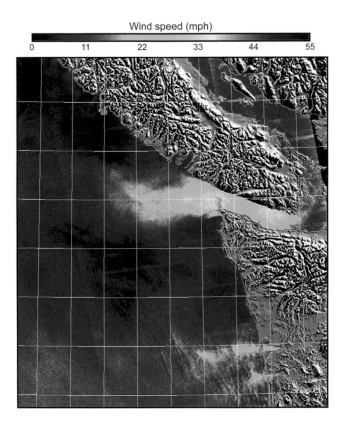

7.10. Surface winds over the Strait of Juan de Fuca and the nearby Pacific Ocean based on satellite measurements using radar. Highest wind speeds are shown by yellow and red, with the largest values at the exit of the Strait near Tatoosh Island. Graphic courtesy of Dr. Nathaniel Winstead.

Westerly Wind Surges through the Strait

Although the easterly gales near Tatoosh Island are perhaps the most well-known wind feature of the Strait, strong winds can also come from the west. During the summer, when high pressure builds offshore and pressure falls over the warm interior of Washington, winds of 20–30 miles per hour can push eastward through the Strait during the afternoon and early evening (see chapter 6 on local sea-breeze winds). Also during the warm season, major marine pushes, in which unusually warm air is replaced by a cool marine flow, can cause wind speeds over the eastern Strait to accelerate to 40–50 miles per hour for several hours. However, there is another type of westerly wind event that can bring even stronger winds to the eastern Strait and the northern Puget Sound: wintertime westerly wind surges associated with certain types of Pacific disturbances that can accelerate winds to as high as 60–90 miles per hour. Such events have damaged well-known harbor restaurants, produced extensive power outages, damaged docks and ferries, and are a real threat to the U.S. Navy homeport along the Everett shoreline. The naval station currently accommodates the USS *Abraham Lincoln* (a *Nimitz*-class aircraft carrier), a destroyer, three frigates, and the Coast Guard's buoy tender. There is little doubt that future westerly surges through the Strait will expose this important facility to hurricane-force gusts.

A particularly strong westerly strait surge occurred on December 17, 1990. As a well-defined upper-level disturbance, known as an upper trough, approached northern Washington from the northwest, and an associated front moved through the Strait, strong winds pushed into the central and eastern Strait. Over a period of a few hours, sustained winds of 50–70 miles per hour, with gusts to 70–80 miles per hour, extended across Whidbey Island into Snohomish County, causing considerable tree falls and loss of power. By the end of the storm, nearly 141,000 Snohomish and Island county customers were in the dark and many roads were impassable. According to the Snohomish County Public Utility District, the December 17 event was the "worst in history from the standpoint of disruption, hardship, and dollar damage," eclipsing the destruction of even the great Columbus Day Storm of 1962. Winds were particularly severe in the Everett harbor, where some ship anemometers measured 70- to 80-mile-per-hour winds. That day, the Washington State ferry *Elwha* (figure 7.11) was docked for repairs at the Fisherman's Boat Shop on the Everett waterfront. As the winds rose above 50 miles per hour, the ferry was thrown against the pier by increasing wave action, destroying a wooden barrier and snapping off pilings. Subsequently, the ferry beat against a concrete pier until the vessel's car deck crumbled. After four hours of damage, the *Elwha*

7.11. The Washington State ferry *Elwha* after repair. The ferry was seriously damaged during the Strait of Juan de Fuca westerly wind surge of December 17, 1990. Photo courtesy of Steven J. Pickens.

was saved by two tugs that pulled the boat into open water. Wind-driven waves also destroyed one of the wing walls at the Lopez Island dock in the San Juan Islands that night.

A more recent westerly surge through the Strait occurred on October 28, 2003, and resulted in extensive tree damage, power outages for nearly a hundred thousand customers, one death, and the loss of Ivar's seafood restaurant in Mukilteo. This storm has much in common with the December 17, 1990, event: a sharp, upper-level trough embedded in northwesterly flow aloft approached Washington State. As an associated Pacific front passed across the Strait, winds increased to over 60 miles per hour across Whidbey Island, nearby Smith Island, and Snohomish County, and large wind-driven

7.12. A large mural in the foyer of Ivar's Mukilteo Landing restaurant suggests that the restaurant's destruction was due to a "rogue" wave (with some lightning thrown in for good measure). In reality, the facility was destroyed by persistent wind-driven waves forced by a westerly wind surge through the Strait of Juan de Fuca. Painting by Chris Hopkins and reprinted courtesy of Bob Donegan and Ivar's Inc.

waves struck the eastern shore of Puget Sound. Such waves pushed into and under Ivar's Mukilteo Landing restaurant, located on the water near a Washington State ferry terminal. This resulted in an abrupt end to the meals of several patrons and the destruction of the restaurant (figure 7.12). The restaurant's signature wood carp was swept away during the storm, but was found two weeks later

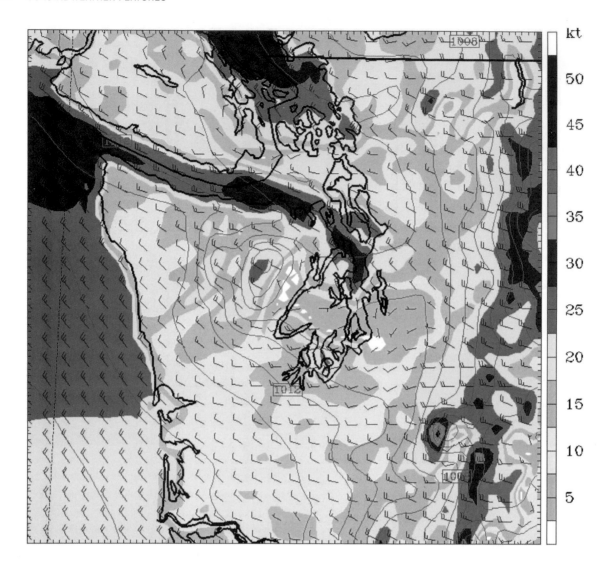

7.13. Twenty-four-hour high-resolution computer forecast valid at 4:00 PM on October 28, 2003, showing predicted surface winds over western Washington. Blue areas indicate sustained winds of 27.5 to 32.5 knots (32 to 37 mph); gusts would be higher. Note the strong winds over the eastern Strait that pushed into northern Puget Sound.

after a substantial reward (a year's worth of fish and chips) was posted. An important aspect of the October 28 event was how well it was predicted by the latest generation of high-resolution computer forecasting models: as illustrated in figure

7.13, the MM5 model, run daily at the University of Washington, was able to skillfully predict the westerly surge a day ahead.

What is the origin of westerly surges in the Strait of Juan de Fuca? Most evolve in a similar way, with a sharp intense upper-level trough that is embedded in a strong jet stream from the northwest. Figure 7.14a shows an example of the conditions for 4:00 PM on December 17, 1990, at a pressure of 500 millibars, a level located approximately 18,000 feet above sea level. This chart is like a topographic

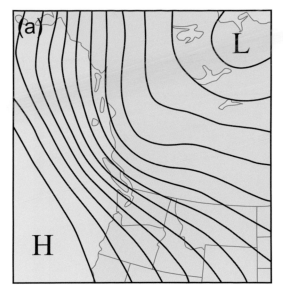

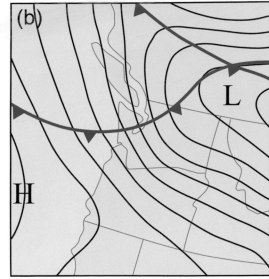

map, with the solid lines telling how high above sea level one must ascend for the pressure to drop to 500 millibars. High heights indicate where this pressure surface tilts upward, while areas of lower heights (known as troughs) are regions where this pressure is found at lower elevations. Winds tend to parallel these lines, and the strength of the winds is proportional to the distance between the lines (closer lines, stronger winds). A well-defined trough is found over southern British Columbia and is indicated by curved, closely spaced lines. A ribbon of tightly packed lines extends from Alaska into the Northwest: this represents the jet stream, in which the trough is embedded. At the same time, a Pacific front, which was associated with the trough, was just about to enter Washington State (figure 7.14b). The wind shifted to the west and northwest behind the front, and because of the orientation of the upper trough, the winds aloft were not only strong, but were also directed eastward down the Strait. Behind the front, pressure increased rapidly, contributing to a pressure difference along the Strait that accelerated winds to the east. Thus, with strong winds directed eastward down the Strait throughout the lower atmosphere

and a large, favorable pressure difference across the gap, the situation was ideal for producing a strong westerly wind surge.

Strong Southeasterlies over the Eastern Strait

Strong southeasterly winds can affect the eastern Strait of Juan de Fuca when low-pressure systems and associated strong southerly winds approach the coast. During such events, powerful and sustained southeasterly winds of 20–40 miles per hour, with higher gusts, occur over the eastern strait and extend eastward to northern Whidbey Island and

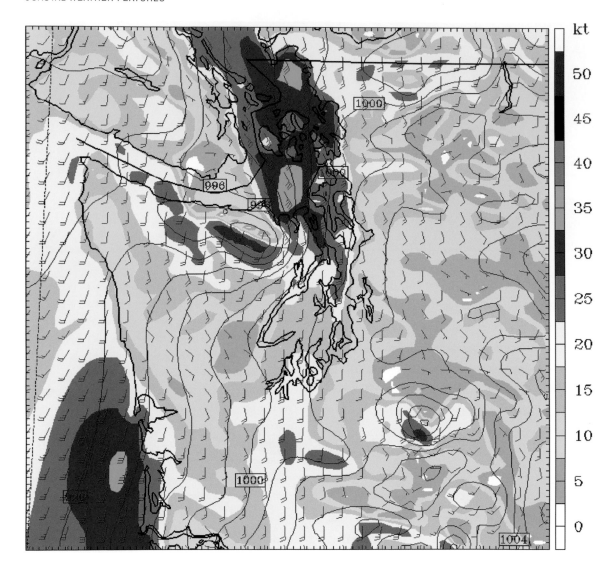

7.15. High-resolution prediction of near-surface wind speed (shading), wind direction (red wind flags pointing in the direction to which the wind is blowing), and sea-level pressure (black lines, millibars) valid at 4:00 PM on January 31, 2006. Southeast winds exceeded 28 knots (32 mph) over the eastern Strait of Juan de Fuca and the Strait of Georgia. Image from the regional model system of the Department of Atmospheric Sciences, University of Washington.

northward into the Strait of Georgia. An example of such strong winds occurred on January 31, 2006, around 4:00 PM when southeast winds gusted to 56 miles per hour at Smith Island, 43 miles per hour at Whidbey Island Air Station, 47 miles per hour at Friday Harbor, and 60 miles per hour at a buoy on the east side of the Strait. A high-resolution computer forecast of the winds and pressure at this time skillfully predicted the strong winds and illustrated how they can extend into the Strait of Georgia (figure 7.15).

To understand why strong southeasterly winds often occur over the eastern strait, one must appreciate the effects of the Olympic Mountains. When the incoming winds in the lower atmosphere are from a southerly direction, air rises on the southern slopes of the Olympics and then descends on the northern side. Rising air cools and becomes denser, resulting in high pressure on the southern side of the barrier, while sinking air on the northern side warms as it descends, causing pressure to fall (see figure 7.15). As a result, the Olympics can greatly modify the regional sea-level pressure distribution, producing a windward zone of high pressure (the windward ridge) and leeward zone of low pressure (the lee trough). Between the windward ridge and the lee trough, pressure differences are enhanced, particularly along the central Washington coast and the eastern Strait of Juan de Fuca. Such large pressure changes result in strong winds. The strength of the mountain-induced pressure pattern and the resulting winds are enhanced as the incoming southerly wind increases, often when a Pacific low-pressure area approaches the coast.

A good example of this effect was evident at 1:00 PM on March 5, 1988, when southerly winds of 30–40 miles per hour approached the Olympic crest as a strong Pacific front and associated low pressure neared the coast. During this period there was a special weather experiment underway in which extra barometers were placed around the Olympics, making possible a detailed pressure analysis (figure 7.16). This analysis shows the windward high-pressure area on the southern side of the barrier, the leeward trough on the northern side of the Olympics, and the area of large pressure difference and strong winds between them. In con-

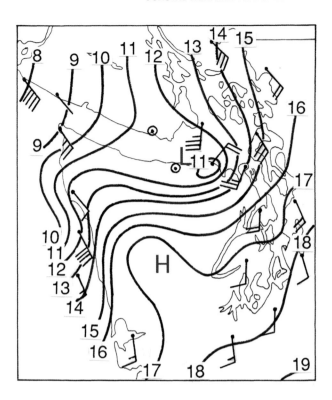

7.16. Sea-level pressure and surface winds for 1:00 PM on March 5, 1988. Air moving up and over the Olympics resulted in a typical pressure variation around the mountains. A windward area of high pressure south of the Olympics ("H") and a leeward low-pressure trough to its north ("L") produced areas where pressure changed rapidly and winds were strong. Graphic from Mass and Ferber (1990), courtesy of the American Meteorological Society.

trast, differences of pressure were small within the lee trough and windward ridge, and thus the winds were weak there. The contrasts in wind over small distances were profound: while gale-force winds were blowing over northern Whidbey Island and nearby waters, calm winds covered Port Angeles and vicinity. This calm-wind zone is often found over Sequim, which may explain why the Native American word for the location means "place of tranquil waters."

The Strait of Georgia and Fraser River Northerlies

Although the strongest winds over the eastern Strait of Juan de Fuca and the inland waters of northwest Washington are generally from the west or southeast, powerful winds occasionally come out of the north. One situation that produces moderate-strength northerlies (20–40 miles per hour) is when high pressure builds to the north and northwest while low pressure is located over southern Washington and Oregon. Air then accelerates southward down the Strait of Georgia, producing strong winds over the water and adjacent land.

However, the most extreme northerly winds have another cause: strong outflow from the Fraser River valley. As described in chapter 4, the Fraser River valley represents a gap in the Cascades and its extension into British Columbia, the Coast Mountains. When Arctic high pressure moves into British Columbia and relatively low pressure builds over Washington, air accelerates southwestward down the Fraser Valley. This northeasterly flow can gust to 60–80 miles per hour at times and has reached 100 miles per hour for the most extreme events.

An example of such a strong Fraser outflow wind occurred on December 28, 1990. A frigid high-pressure center had moved into British Columbia, while a trough of low pressure associated with an Arctic Front stretched from Wyoming, through Washington, and offshore of Vancouver Island (figure 7.17). The result was a zone of intense pressure difference over southern British Columbia and northern Washington. This pressure difference accelerated cold air southwestward down the Fraser River gap and then across Bellingham, the

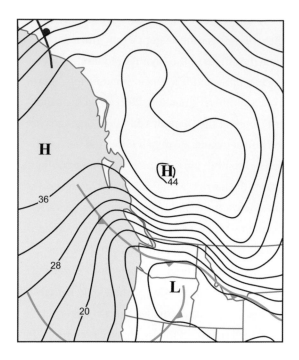

7.17. National Weather Service surface-weather chart for 4:00 AM on December 28, 1990. The solid lines are isobars (lines of constant pressure). A high-pressure area was centered over the interior of British Columbia, with low pressure over southern Washington and Oregon. The zone of large pressure difference between these two features produced strong northeasterly winds in and downwind of the Fraser River valley. Illustration by Beth Tully/Tully Graphics.

San Juan Islands, and vicinity. As shown in figure 7.18, a swath of extremely strong winds, reaching 80–90 miles per hour, struck a large area of northwest Washington. Two miles north of Bellingham Airport, a gust of 115 miles per hour destroyed a truck storage shed, while in Bellingham harbor two barges were torn from their moorings and sent adrift. Also in that harbor, a Georgia Pacific boiler stack was snapped off 30 feet from the top. The northeasterly air current maintained its

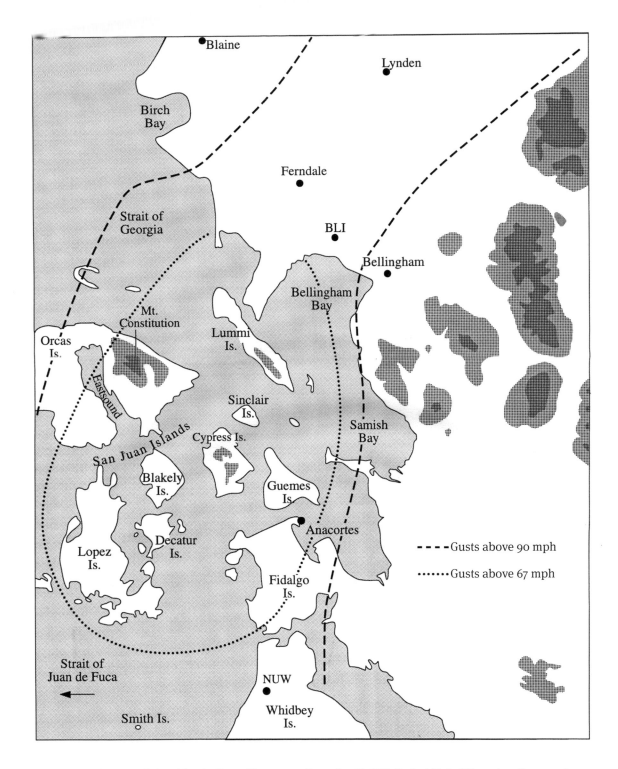

7.18. The area of strong winds exiting the Fraser River gap on December 28, 1990. Dashed (dotted) lines show the area where winds exceeded 67 (90) miles per hour. Terrain is indicated by dark shading. BLI is Bellingham Airport; NUW is Whidbey Island Naval Air Station. Graphic courtesy of the American Meteorological Society.

7.19. Dozens of boats and docks were destroyed or damaged around Anacortes, Washington, on the northern shore of Fidalgo Island during the December 28, 1990, Fraser Gap windstorm. This view faces north to Guemes Island. Photo courtesy of Scott Terrell, *Skagit Valley Herald*.

integrity and strengthened as it passed over the water, since water offers less drag than land. On northern Lummi Island, two observers reported winds exceeding 100 miles per hour, and massive tree falls, damage to homes, and extensive power outages occurred. A few miles to the south, Guemes Island was hit hard: there was a widespread loss of trees over the northern portion of the island, with swaths roughly 600 feet long and 150 feet wide where nearly all trees were downed. Strong winds extended even farther south, with a gust at Anacortes on Fidalgo Island measured

at 100 miles per hour and several dozen boats and a similar number of boathouses destroyed (figure 7.19). Immediately south of Fidalgo Island, Deception Pass State Park on northern Whidbey Island experienced "the greatest devastation the park ever had," according to a park spokesperson, with motorists in the park trapped by downed trees.

Strong northeasterly winds exiting the Fraser River gap have repeatedly struck the San Juan Islands, with particularly significant effects on the higher portions of Orcas Island. Even a casual hike near the top of Mount Constitution, a 2,409-foot peak in Moran State Park on the eastern side of the island, reveals numerous downed trees, with most falling to the southwest. A particularly savage Fraser River windstorm devastated the park in 1972, with a huge loss of trees and many blocked roads.

THE PUGET SOUND CONVERGENCE ZONE

One or two dozen times a year, a band of clouds and precipitation settles in over the central Puget Sound, with nearly cloud-free skies to the north and south. During these periods, northern Seattle and the suburbs stretching to Everett can be cool and rainy, while a dozen or so miles to the south the sun is shining and temperatures are 5–10 °F warmer. This important local weather phenomenon, called the *Puget Sound Convergence Zone*, is now well understood and often skillfully forecast.

For a number of years forecasters had noticed that clouds and precipitation often remained over central Puget Sound following the passage of Pacific fronts, with some forecasters even suggesting that "secondary fronts" were lurking in the Pacific. By the 1970s, the true nature of this phenomenon became evident: a mountain-induced zone of converging air over Puget Sound. During convergence zone events, low-level winds over the coastal waters of Washington State are typically from the west to northwest. Such a wind direction often occurs after frontal passage, particularly during the spring and early summer. At low levels, this northwesterly flow is blocked by the Olympic Mountains and deflected through the Strait of Juan de Fuca and Chehalis gaps, located north and south of the Olympics, respectively (figure 7.20a). Blocked by the Cascades, these airstreams are forced to converge over the central and northern Puget Sound. Converging air at low levels forces some air to rise, which in turn produces a band of cloud and precipitation across the Puget Sound basin (figure 7.20b) that is often located between Seattle and Everett.

Figure 7.21 presents satellite and weather-radar imagery for a typical convergence zone event. The satellite picture not only shows an east-west cloud band across Puget Sound, but clear zones to the north and south (figure 7.21a). Such clear areas frequently accompany convergence zone events and are associated with downward air motion to the north and south of the convergence zone, produced by air descending the western slopes of the Olympics and the mountains of Vancouver Island. The satellite imagery indicates that the air off the Pacific is moist and unstable, with enhanced clouds where air is forced to rise on the windward side of mountains or in the convergence zone. A second weak convergence zone extends from Victoria, Canada, eastward over the San Juan Islands, as southwesterly air originating from the Strait of Juan de Fuca meets northerly air moving southward down the Strait of Georgia. Radar imagery, which delineates the location and intensity of precipitation, shows that the convergence zone precipitation is limited to a relatively narrow band of moderate and heavy showers between Seattle and Everett (figure 7.21b). Showers are also found on the western (windward) side of the Cascades, as well as in the Strait of Georgia convergence zone.

During the May 5, 2002, convergence zone, the weather contrasts over Puget Sound were substantial. For example, at the time of the imagery in figure 7.21, Paine Field in the north Sound had completely overcast skies, a temperature of 42 °F, and rain showers. Meanwhile, Seattle-Tacoma Airport enjoyed partly cloudy skies, sunshine, and a temperature of 51 °F. Since convergence zone events happen a dozen or so times each year, knowing their ways can greatly improve one's chances for enjoying dry and even sunny outdoor recreation.

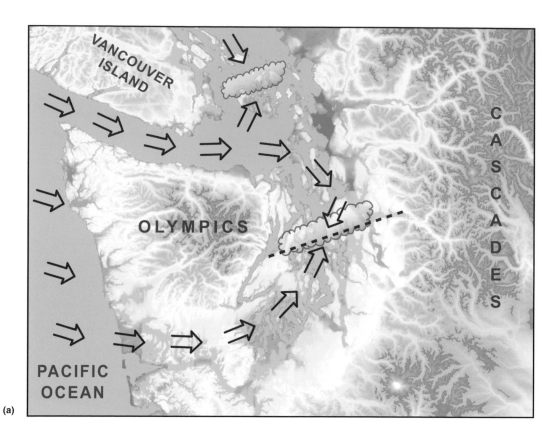

(a)

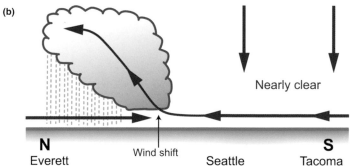

(b)

Nearly clear

N
Everett

Wind shift

Seattle

S
Tacoma

7.20. (a) A Puget Sound Convergence Zone often forms when the flow on the coast is from the west to northwest. Low-level air passes around the Olympic Mountains and then converges over Puget Sound. This forces some air to rise, forming an east-west band of clouds and precipitation extending from the Olympics to the Cascades. Arrows indicate low-level winds, the red dashed line gives the position of the surface convergence line, and the typical locations of clouds and precipitation are indicated. (b) Vertical cross-section through a convergence zone. South of the convergence, there are often clear skies produced by air sinking off the Olympics. Illustrations by Beth Tully/Tully Graphics.

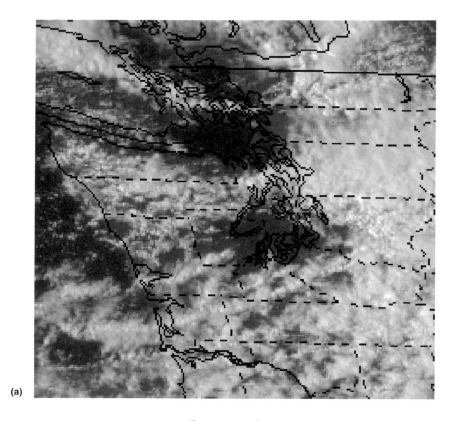

(a)

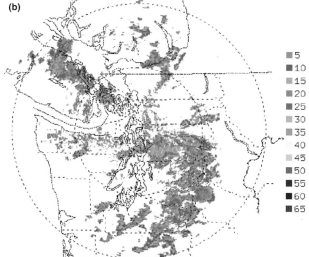

(b)

■	5
■	10
■	15
■	20
■	25
■	30
■	35
	40
■	45
■	50
■	55
■	60
■	65

7.21. (a) High-resolution National Weather Service weather-satellite photo at 4:45 PM PDT on May 5, 2002. The band of clouds stretching across Puget Sound is associated with an active Puget Sound Convergence Zone. (b) National Weather Service weather-radar image from its Camano Island facility at 4:59 PM PDT on May 5, 2002. A band of moderate to heavy showers produced by the Puget Sound Convergence Zone extends across northern Puget Sound. Other showers are found along the western slopes of the Cascades. Heaviest showers are indicated by green and yellow colors.

Convergence zone clouds and precipitation are generally found over the central and northern sound, but can push as far south as Tacoma or stray north across central Whidbey Island.

Convergence zones are most frequent in the spring and early summer when coastal winds are more often from the west to northwest and the atmosphere is still moist enough for clouds and precipitation. The spring is also the period when air over the region is least stable (when temperature drops more rapidly with height), and convective clouds (cumulus and cumulonimbus) can be initiated by the upward motion associated with the convergence zone. On very few occasions, convergence zone–related thunderstorms have spawned weak tornadoes, while during the cool season, convergence zone snowstorms have occurred. As discussed in chapter 4, a convergence zone snowstorm on December 18, 1990, dropped 10–12 inches over central and north Seattle, while little fell at Seattle-Tacoma Airport to the south or Everett to the north. The melting of heavy precipitation within a convergence zone can help cool the air sufficiently so that snow can fall even at low elevations.[4] The precipitation in convergence zones often extends into the Cascades, where it can greatly enhance mountain snowfalls. Frequently, convergence zone snow is heavy along US 2 and Stevens Pass, while Snoqualmie Pass, 40 miles to the south, is enjoying fair conditions.

Although convergence zone events can occur any time of day, during the spring and early summer they tend to form near Everett during the late morning and then move southward toward north Seattle during the afternoon. This tendency is probably caused by the typical local winds during the warm season in which northerly winds develop over the north Sound during the afternoon (these winds are described in chapter 6). At night, convergence zones tends to weaken and retreat toward the north.

4   The main mechanism of the cooling is the melting of snow falling out of clouds above. The more intense the precipitation, the more melting snow falls into the warmer layer below, progressively cooling the layer so that eventually snow can reach the surface.

8

MOUNTAIN-RELATED WEATHER PHENOMENA

AS NOTED THROUGHOUT THIS BOOK, NORTHWEST WEATHER IS PROFOUNDLY

influenced by the region's mountains. Precipitation is greatly enhanced on the windward side

of terrain, but reduced in the lee. Mountains isolate the influence of the temperate Pacific to

mainly the coastal half of the region, while gaps in terrain provide conduits for cold air or can

produce localized areas of strong winds. However, the influence of Northwest mountains is

not limited to these examples, and this chapter examines four fascinating mountain-related

weather features of the Northwest: mountain waves forced by air being pushed upward by

terrain, which produce clouds reminiscent of flying saucers and are the stuff glider pilots

dream of; extreme downslope windstorms that have struck foothill towns such as Enumclaw,

8.1. A mountain wave cloud over Mount Rainier. The resemblance to a flying saucer is evident. Photo taken by meteorologist Art Rangno from a University of Washington research aircraft.

Washington; the profound effects of the only sea-level passage through the Cascade Mountains, the Columbia River Gorge; and the frequent avalanches that strike Northwest slopes.

MOUNTAIN WAVES AND FLYING SAUCERS

On June 24, 1947, Kenneth Arnold, a businessman from Boise, Idaho, was flying a small plane near Mount Rainier when, according to Associated Press reports, he spotted a chain of nine "saucer-like" objects above and east of the mountain. Brilliant in the sun, these objects darted toward Mount Adams at "an incredible speed" that he estimated to be at least 1,200 miles per hour. Arnold's story of saucer-shaped objects initiated a UFO craze that has not abated. Analyses by meteorologists and other scientists suggest that Mr. Arnold did not spot a visitor from another world, but rather a *mountain wave*

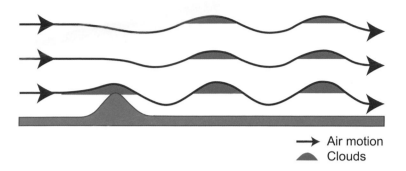

8.2. Schematic of wave cloud formation caused by air forced to rise over a mountain barrier. Clouds form where air is pushed upward and dissipate when the air sinks. The mountain can initiate a vertical oscillation that can continue for many miles downstream. Illustration by Beth Tully/Tully Graphics.

→ Air motion
▲ Clouds

cloud, a frequent visitor to the mountainous Pacific Northwest (figure 8.1).

What produces mountain waves? Most of the earth's atmosphere is stable, which means that when a parcel of air is displaced up or down it returns to its original position. More exactly, if stable air is pushed up or down it will return to its original position after going through a few up and down oscillations. A good analog is a pendulum or a swing. Given a push, both will move back and forth a few times before stopping in their original position.

When stable air approaches a mountain barrier, it is forced upward on the windward side of the mountain and then descends in the lee (figure 8.2). Often the air will continue to oscillate in the vertical after it passes the mountain crest, producing a series of mountain lee waves. These kinds of mountain lee waves are called trapped lee waves, because they remain trapped in the lower to middle atmosphere. Such waves are often associated with winds increasing with height. If the air is relatively moist, the upward motion forced by the mountain or by the subsequent waves will cool the air sufficiently to cause saturation and form clouds, since cool air can hold less water vapor than warm air (see chapter 2 for a discussion of why clouds form). The cloud that forms right on the mountain crest is called a cap cloud and the downwind clouds are

known as lee wave clouds. Sometimes only the cap cloud is evident, while on other occasions several lee wave clouds occur downstream. When a mountain crest extends for tens of miles or more, the lee wave clouds are often evident on satellite photos as a series of parallel lines downwind of the mountain crest (figure 8.3a). In contrast, a large isolated barrier, such as Mount Rainier, can produce a V-shaped cloud configuration, reminiscent of the wake of a ship (figure 8.3b). Both types of lee waves can extend dozens to over 100 miles downwind of the mountain barrier.

Mountain wave clouds, particularly those produced by relatively isolated barriers, are often lens shaped and thus described by the Latin word *lenticularis*. Such lens-shaped clouds (like the one in figure 8.1) usually reside at midlevels (5,000–20,000 feet) and are known as altocumulus lenticularis, since altocumulus clouds are middle-level clouds that are broken into identifiable elements. When the mountain barrier or the incoming flow becomes more complex, mountain wave clouds may no longer look like perfect disks, but they still maintain an other-worldly appearance. For example, when atmospheric moisture has some layering, with alternating dry and moist layers, a stack of mountain wave clouds can appear (figure 8.4).

If the incoming flow aloft is steady, mountain

(a)

(b)

8.3. (a) NASA MODIS AQUA satellite image of north-central Washington State on July 13, 2006. Such long lines of lee wave clouds are produced by extensive mountain barriers. (b) NASA MODIS AQUA satellite image of central Washington on November 8, 2004. The V-shaped mountain wave clouds were forced by Mount Rainier.

wave clouds will generally remain stationary, even though air is continuously moving through them. However, if the flow aloft is altered, perhaps due to an incoming storm system, the locations of upward motion, and thus the wave clouds, can move rapidly, producing the extraordinary swift motion that suggested an alien visitation to Kenneth Arnold in 1947.

In the Northwest, the appearance of mountain cap clouds or wave clouds is often the first sign of an approaching storm. The formation of such clouds usually requires at least a moderate wind speed (20–30 miles per hour) for the air approaching the mountain, as well as high relative humidity. Such conditions can be produced by approaching weather systems, which not only increase wind speed but also produce gentle upward motions that cool the air and thus increase relative humidity. Mountain waves are frequently observed over and east of the Cascades, since regional winds

8.4. Mountain wave clouds downwind of Mount Adams. Note the multilevel cloud structure, reminiscent of stacked plates. Photo courtesy of Darlisa May Black.

aloft are generally from the west. When easterly or southeasterly flow occurs, mountain wave clouds can occur over the western slopes of the Cascades and even over Puget Sound and the Willamette Valley. Perhaps an example of such wave clouds inspired Ohio congressman Dennis Kucinich to admit during the Democratic presidential debate on October 30, 2007, that he had seen a UFO during a visit to Washington State. His location at the time, 30 miles northwest of Mount Rainier in Graham, Washington, suggests that, like many before, Congressman Kucinich had mistaken a mountain-produced lenticular cloud for a visitor from another world.

Northwest glider pilots have a considerable interest in local mountain waves, since the upward motion in the waves can be used to gain and maintain altitude. The so-called Wenatchee Wave, named after the mountain wave produced by the ridge west of Wenatchee, Washington, is particularly popular. Strong mountain waves are also produced by the large Cascade volcanoes and have allowed gliders to gain great altitude. For example, one local glider pilot, Vitek Siroky, used the Mount Rainier mountain wave to reach an elevation of over 27,000 feet on September 28, 1997.

DOWNSLOPE WINDSTORMS OF THE PACIFIC NORTHWEST

Although the weather of the lower western slopes of the Cascades is generally benign, there is one region that experiences 50- to 80-mile-per-hour winds roughly five times a year and occasionally is buffeted by winds exceeding 100 miles per hour. In this area, stretching along the western Cascade foothills near the town of Enumclaw, Washington, hurricane-force winds are a regular visitor, resulting in special building codes and residents prepared for frequent power outages. As noted in chapter 5, these windstorms occur downstream of a low area in the Cascades that allows air from eastern Washington to push across the mountains and to accelerate as it descends into the Puget Sound lowlands. But there is much more to the story.

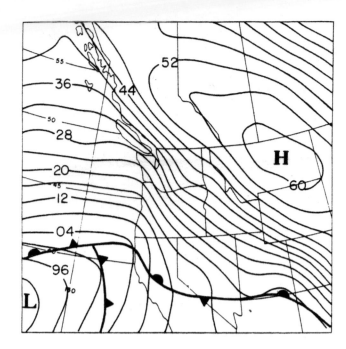

8.5. National Weather Service sea-level pressure analysis for 4:00 AM on December 24, 1983. With intense high pressure over the interior and a low-pressure area approaching the coast, a large pressure difference developed over Washington and Oregon, resulting in intense winds in Enumclaw and surrounding communities. Black lines are lines of constant pressure (where "20" indicates 1,020 millibars).

Probably the greatest downslope windstorm of the past century in the Pacific Northwest struck the western Cascade foothills on December 24, 1983. During the prior few days, a frigid Arctic air mass had spread over the region, with single-digit temperatures west of the Cascades and subzero temperatures over eastern Oregon and Washington. Associated with the cold air, extraordinarily strong high pressure dominated the interior Northwest, including the highest value ever experienced over the continental United States (1,064 millibars, or 31.42 inches of mercury, at Miles City, Montana). With high pressure inland and low pressure approaching the Northwest coast (figure 8.5), an intense pressure difference developed

(a)

(b)

8.6. Hurricane-force downslope winds produced massive damage in Enumclaw, Washington, and the surrounding area on December 24, 1983. (a) A number of mobile homes as well as three high-tension towers (a crumbled tower is seen in the right background) were destroyed by winds exceeding 100 miles per hour. (b) Strong winds stripped the roofs off dozens of barns and farm buildings.

across the Washington and Oregon Cascades. This large pressure difference, combined with moderate (20–40 miles per hour) southeasterly winds at mountain-crest level, pushed air eastward across the Cascades. In the narrow Columbia River gorge, easterly winds gusted to 50 miles per hour, producing significant damage and power outages. But it was along the western Cascade foothills, from Enumclaw toward Black Diamond, that the strongest winds and most devastating damage would occur (figure 8.8).

During the late evening and early morning hours of December 23–24, the pressure difference across the Cascades rose rapidly (to a huge 17.3

millibars, .51 inches of mercury) as the low pressure approached the coast, causing the easterly winds to intensify. By 4:00 AM, sustained winds of 60–80 miles per hour, with gusts over 100 miles per hour, were observed at several locations near Enumclaw. In the Veazey neighborhood, 3 miles northeast of downtown Enumclaw, winds reached 118 miles per hour before the anemometer was blown off the roof, while at the City of Enumclaw maintenance shops in the northeast part of the city there were sustained winds of 70–85 miles per hour, gusting to 120 miles per hour. The resulting damage was extraordinary: mobile homes were torn from their foundations and tumbled hundreds of feet, downed trees caused a massive power outage and made most major roads impassable, huge high-tension power-line towers were crumpled, and sheet-metal roofs were torn off area barns (figure 8.6). Strong winds pushed westward toward Puget Sound in a well-defined current, about 20 miles wide (figure 8.7). These winds, still gusting to 50–60 miles per hour by the time they reached Puget Sound, caused tree failures and power outages throughout the area (figure 8.8). Wind-driven waves on the Sound badly damaged docks as well as the Point Defiance boathouse in Tacoma. The current of strong winds had very distinct boundaries; for example, twenty miles to the north of Tacoma in Seattle the winds were light, and light winds were found a short distance to the south of the maximum winds as well. Strong winds were observed at the Crystal Mountain ski area (6,800 feet), with gusts reaching 116 miles per hour. Lesser, but still damaging, winds were found west of Stevens Pass and Snoqualmie Pass. An interesting aspect of the damage from large downslope windstorm events such as December 1983 is that

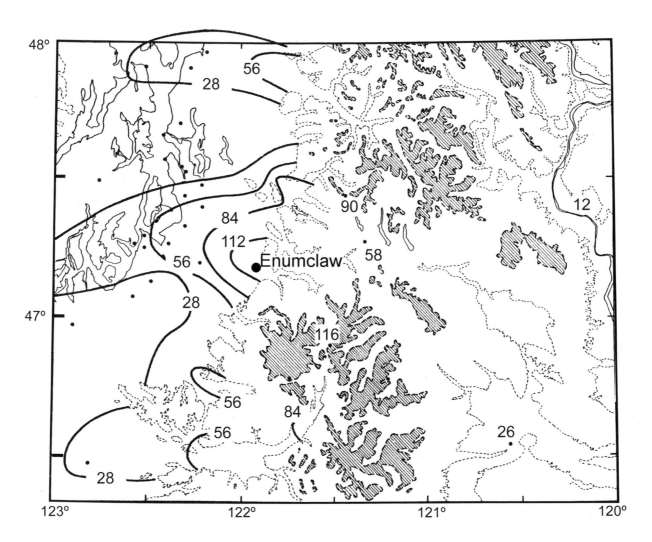

8.7. Maximum winds (mph) observed on the morning of December 24, 1983. Note that the strongest lowland winds were near the foothills town of Enumclaw, Washington, and that a current of strong winds extended eastward toward Puget Sound. A lesser current to the north was associated with Stevens Pass. Graphic from Mass and Albright (1985).

the strongest winds are not uniform; rather, the greatest damage is usually concentrated in narrow swaths a few hundred feet wide.

Why are the strongest winds limited to the Cascade foothills near Enumclaw? The reason is made clear from a topographic map of the region (figure 8.9). The Cascade Mountains of Washington are not of uniform height in the north-south direction. The highest terrain is found over the northern Cascades and over the southern Cascades from Mount Rainier to the Columbia Gorge. Between these two sections is the relatively low Stampede

Gap region, named after Stampede Pass. When air approaches the Cascades from the east, it naturally passes through the lowest sections first. The lowest, of course, is the near sea-level Columbia River gorge, but that channel is too narrow to allow a

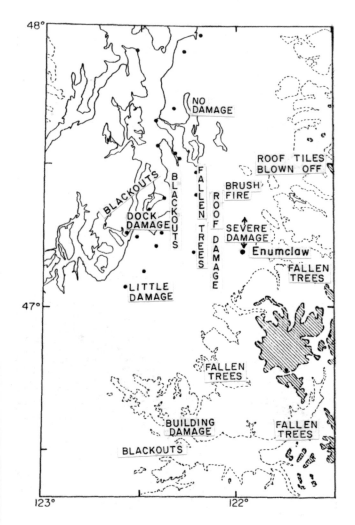

8.8. Damage reports for the December 24, 1983, windstorm. Graphic from Mass and Albright (1985).

similar to constant-level gap winds (described in chapter 7), such as through the Strait of Juan de Fuca or the Columbia Gorge, with air accelerating as it moves within the gap from higher to lower pressure. But in addition, there are other mechanisms in play. One is analogous to the acceleration of cold, dense air down a slope or of a ball picking up speed as it rolls downhill. Another mechanism is connected with the mountain waves discussed earlier in the chapter. Under the proper circumstances, which were fulfilled during the December 24 wind event, the mountain wave, produced by the flow passing over the Cascades, amplifies, producing an additional acceleration down the mountain slopes. On December 24, 1983, all these mechanisms came together to produce a storm unequalled in the modern observational record or the memories of long-lived inhabitants of the area.

A more recent "Enumclaw-style" windstorm on December 5, 2003, brought wind gusts reaching 80 miles per hour from Enumclaw northward to Maple Valley, closing dozens of roads with fallen trees and knocking out power to more than two hundred thousand homes. Many hiking trails along the western slopes of the Cascades were covered with fallen trees and two people were sent to the hospital with head injuries after a 110-foot-tall tree crashed on their Ford Explorer. According to a University of Washington study, the name *Enumclaw* is derived from a Salish word meaning

large volume of air to pass westward. Next lowest is the Stampede Gap region, which is tens of miles wide and riddled with passes of only 3,000–4,000 feet in elevation. This weakness in the mountains allows a large quantity of air to move across the Cascades, particularly when deep, cold air is entrenched over eastern Washington.

The physics of the acceleration across and down Stampede Gap is complex. Part of the speed-up is

(opposite)

8.9. Regional- and local-scale topographic maps of the December 24, 1983, windstorm region. Windy locations such as Enumclaw, Washington, are west of Stampede Gap, a major weakness of the central Cascade Mountains. Graphics from Mass and Albright (1985).

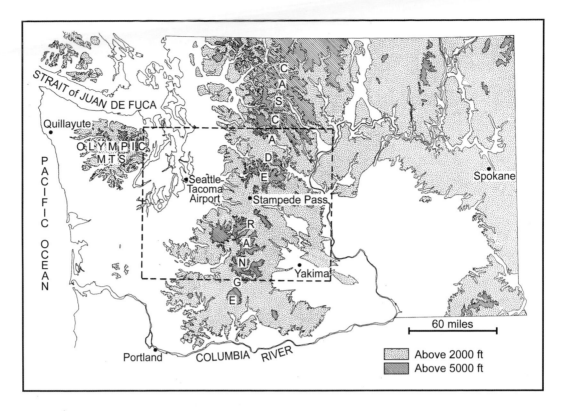

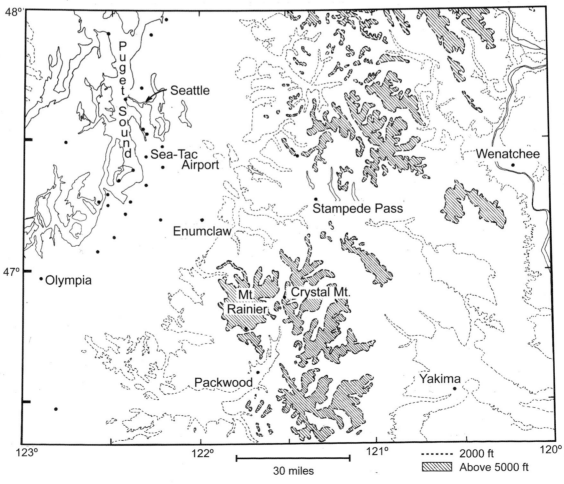

(a)

(b)

8.10. The westerly downslope windstorm of January 7, 2007, produced extensive damage in and near Wenatchee, Washington, where many roofs lost their shingles and trees were toppled throughout the area. (a) Aerial view of the damage. Photo courtesy of the Chelan County Sheriff's Office, Michael T. Harum, sheriff. (b) Lines of trees were toppled in the city's Lincoln Park, with substantial damage to fences and structures. Photo courtesy of Amber Barber.

"place of evil spirits." Clearly, the early residents recognized the dangerous and often terrifying aspects of this windblown locale.

Flying into Seattle-Tacoma Airport on an Enumclaw windstorm day is a long-remembered experience. Descending across the Cascades, one often experiences moderate to severe turbulence over the Cascade slopes, and moderate turbulence within the strong current of easterly winds at low levels. The narrow extent of the windstorm current is clearly apparent from the air, with substantial whitecaps and waves on Puget Sound associated with the strong winds, and a flat sea to the north and south.

The strong easterly winds associated with outflow from Stampede Gap are often the first of a one-two punch to strike the nearby Cascade foothills and neighboring towns. With high pressure over the interior, a strong low-pressure center can approach the coast, building up a large pressure difference across the Cascades and initiating a powerful easterly wind event. Then, as the low center moves north of Puget Sound, strong *southerly* winds develop as the air moves northward toward the low. Such two-step strong winds occurred during the 1993 Inauguration Day Storm and the 1962 Columbus Day Storm, among others. No wonder building codes call for stronger roofs in Enumclaw and vicinity.

Although the Enumclaw area experiences the most frequent and strongest downslope windstorms of the region, this phenomenon does occur in other Northwest areas. When strong westerly winds hit the Cascades, powerful and damaging winds can descend the eastern slopes from Wenatchee to Yakima, with the Kittitas Valley and environs being particularly "favored." Such a downslope windstorm struck in 1972, overturning rail cars south of Vantage, while in 1983 strong westerly winds gusted to 85 miles per hour at Wanapum Dam, 90 miles per hour at Lake Chelan, and 55 miles per hour at Wenatchee. More recently, damaging westerly winds struck the eastern Cascade slopes near Wenatchee on January 7, 2007, with gusts of 72 miles per hour at the Wenatchee airport, 74 miles per hour at Manson on Lake Chelan, and 137 miles per hour at the nearby

Mission Ridge ski resort. Striking around 10:00 AM that day, strong winds damaged or destroyed hundreds of roofs (figure 8.10a), produced a regional power blackout that affected nearly twenty thousand customers, and caused extensive loss of trees throughout the area (figure 8.10b). As with its recent western-slope cousins, the January 7, 2007, windstorm was well forecast by the latest generation of high-resolution prediction models, enabling the Spokane office of the National Weather Service to provide accurate and timely warnings. Computer models have also suggested that strong downslope winds have struck the northeast slopes of the Olympics, but without weather observations on the forested slopes the only evidence of their occurrence is regions of substantial tree falls.

THE WEATHER OF THE COLUMBIA RIVER GORGE

The Columbia Gorge plays a unique role in Northwest weather, serving as the only sea-level passage across the Cascades (figure 8.11). During the summer, as high pressure builds over the eastern Pacific and pressure falls over the heated interiors of eastern Washington and Oregon, air accelerates eastward through the Gorge. As noted in chapter 6, the summer westerly winds over the central and eastern Gorge, coupled with the flow of the Columbia toward the west, produce ideal conditions for world-class wind surfing. In contrast, during the winter, lower pressure over the stormy Pacific Ocean and higher pressure over the cold continental interior results in winds from the east that can bring cold air to the western Gorge and the Portland metropolitan area, sometimes resulting in snow and ice storms (see chapter 4).

Although the Gorge serves as a unique conduit between the east and west sides of the Cascades, its impact on Northwest weather, although substantial, is limited by its narrow width, ranging from approximately 3–8 miles (figure 8.12).

The profound effect of the Gorge on Portland's wintertime weather is illustrated in figure 8.13.

During the cold season, the most frequent winds at the Portland Airport are from the east-southeast, directly out of the Gorge. In contrast, the typical winter winds at Salem, 50 miles to the south, are from the south, similar to most stations in the central and southern Willamette Valley that are away from the influence of the Gorge. Easterly flow from

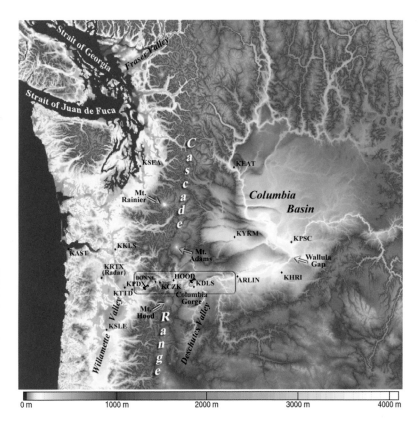

8.11. Topographic map of Washington and northern Oregon. The Columbia River gorge area (outlined by the rectangle) is the only near sea-level gap through the Cascades, allowing the movement of air across the mountain barrier. Indicated locations include Portland (KPDX), Troutdale (KTTD), Cascade Locks (KCZK), Hood River (HOOD), and The Dalles (KDLS). Graphic from Sharp and Mass (2004), courtesy of the American Meteorological Society.

8.12. View of Portland, Mount Hood, and the Columbia Gorge generated with topographic data from a NASA shuttle mission and Landsat satellite imagery. Courtesy of NASA.

the Gorge is the main way for cold air to reach the Portland area, with snow and freezing rain being rare when the winds blow from other directions.

Although westerly winds in the Gorge are most frequent during the summer when warm temperatures and associated low pressure are found over eastern Washington, strong westerly winds sometimes occur over the eastern Gorge during the winter. These westerly gales are usually associated with the eastward passage of Pacific weather systems (generally strong cold fronts) and rarely last longer than a few hours. On December 15, 2000, such a westerly surge produced winds greater than 55 miles per hour over the eastern Gorge and blew a tractor-trailer truck off the Biggs Bridge (figure 8.14), 16 miles east of The Dalles, killing the driver. Today, wind sensors provide warnings of dangerous crosswinds on the bridge deck.

Like the winds of the Strait of Juan de Fuca discussed in chapter 7, the strongest Columbia Gorge winds are in the exit of this gap, not in its center where the Gorge is narrowest. A high-resolution simulation of the surface winds in the Gorge during a typical period of strong easterly

flow is shown in figure 8.15. Although there is a modest wind speed increase in the central Gorge just upstream of Cascade Locks, the greatest wind speeds, exceeding 40 miles per hour, are found at the exit of the gap just to the east of Troutdale. How can we explain this distribution of strong winds? During most strong easterly wind events in the Gorge, there is cold air in eastern Washington banked against the eastern slopes of the Cascades that pushes into the Gorge. This cold air remains relatively deep in the Gorge as long as there are constraining walls. As the gap widens at the western end of the Gorge, the height of the cold air collapses (figure 8.16), just as the level of water would drop after exiting a narrow channel. Since cold air

8.13. Winter weather statistics for the Portland International Airport. (a) Percentage of winter days with various wind directions. (b) Amount of snow associated with different directions. (c) Percentage of days with freezing rain for various wind directions. Strong winds and wintry weather at Portland are mainly associated with days with easterly flow in the Columbia Gorge. For each direction, the distribution by wind speed (knots) is also shown. Graphic from Sharp and Mass (2002), courtesy of the American Meteorological Society.

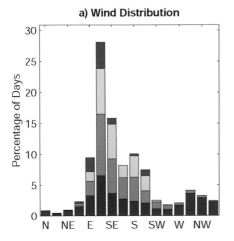

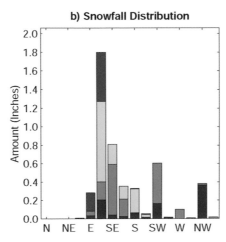

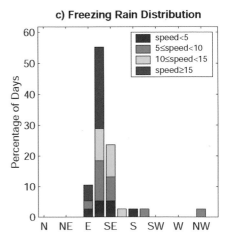

8.14. The Biggs Bridge, located on the eastern side of the Columbia Gorge, sometimes experiences powerful winds from the west. Wind sensors have been installed to provide warnings of strong wind events. Photo courtesy of Jeffrey Elmer.

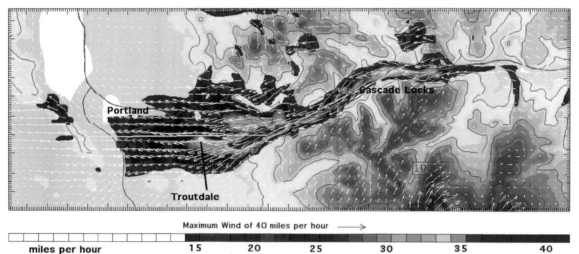

Maximum Wind of 40 miles per hour ⟶

miles per hour   15   20   25   30   35   40

8.15. Simulation of near-surface wind speed from a high-resolution simulation of Columbia Gorge flow. The brown shading is terrain height, and other colors indicate wind speed in miles per hour. The strongest winds (red areas) are over the western Gorge, east of Troutdale, Oregon. Graphic from Sharp and Mass (2002), courtesy of the American Meteorological Society.

is denser and heavier than warm air, the change in cold-air depth at the exit of the Gorge causes a change of pressure at low levels (higher pressure to the east where the cold air is deeper), which accelerates the near-surface air to the west.

The Columbia Gorge has some of the most dramatic precipitation gradients (change of precipitation with distance) in the United States as a result of the changing elevation of the surrounding terrain. Annual precipitation roughly doubles between Portland, near sea level on the western side of the Gorge, and the Bonneville Dam, where average precipitation reaches roughly 79 inches per year (figure 8.17). Precipitation is greatest near Bonneville, which is surrounded by the highest Cascade terrain. Precipitation then falls off by more than half in the 20 miles between Bonneville and

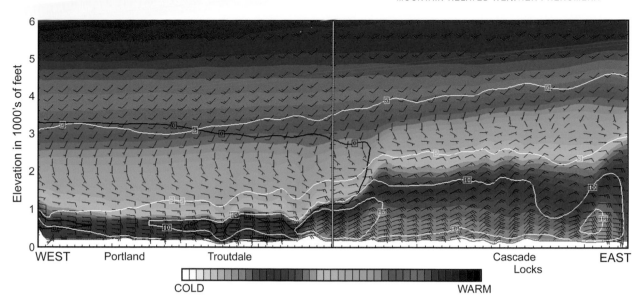

8.16. Simulated temperatures and winds for a vertical cross-section along the eastern half of the Columbia Gorge during a strong easterly wind event. This section extends from sea level to 6,000 feet in the vertical, and from Portland on the far left to just east of Hood River on the far right. The green, blue, and purple colors represent the coldest temperatures. Note how the cold air collapses over the center of the cross-section where the Gorge widens at its exit. Graphic from Sharp and Mass (2002), courtesy of the American Meteorological Society.

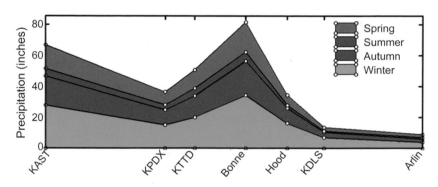

8.17. Annual precipitation (inches) along the Columbia Gorge. Stations include Astoria (KAST), Portland (KPDX), Troutdale (KTTD), Bonneville Dam (Bonne), Hood River (Hood), The Dalles (KDLS), and Arlington (Arlin). Note the large increase in precipitation over the central Gorge and the steep decline to the east. This also shows the amount of precipitation for each season. Graphic from Sharp and Mass (2004), courtesy of the American Meteorological Society.

Hood River, on the lee side of the Cascades, and continues its plunge to desertlike levels at The Dalles, 15 miles farther to the east. Thus, driving on Interstate 84 at highway speeds one can go from a lush forest canopy to desert in roughly a half hour (figure 8.18)

Although the Columbia Gorge is designated a National Scenic Area and thus is afforded protected status, it does experience substantial air pollution. Visibility within the Gorge is often impaired,

and concentrations of sulfur dioxide (burnt match smell), ammonia, mercury, and other pollutants are as high as in major urban areas. There appear to be two primary sources of these pollutants: a coal-fired power plant and a major dairy facility in Boardman, Oregon. The coal burning releases more

(a)

8.18. The large change in precipitation across the Columbia Gorge is evident in the contrasting vegetation along its length. (a) View eastward toward the high terrain of the Gorge from the Crown Point viewpoint. Photo courtesy of Justin Sharp. (b) Picture taken in Wishram, Washington, in the eastern Gorge looking westward toward Mount Hood. Photo courtesy of Barry Jacobson.

(b)

than 20,000 tons of sulfur dioxide every year, and decomposing manure releases thousands of tons of ammonia compounds. The ammonia breaks down into nitrogen compounds, which can form acid rain when they combine with clouds and precipitation.

AVALANCHE WEATHER

Although black ice is probably the worst weather-related killer in the Northwest, avalanches are most likely in second place (figure 8.19). In

8.19. Small avalanche off the east face of Mount Waddington, British Columbia's second-highest peak. Note the substantial windblown cornices on the crest. A slab avalanche appeared to have recently occurred over the lower slopes, as evidenced by the fracture line. Photo courtesy of Peter Neff.

Washington State alone for the period 1910 through 2003, avalanches claimed 204 lives, with the greatest avalanche loss occurring in the Wellington train disaster of 1910 in which ninety-six people perished. Even as this book is being completed during the winter of 2007–08, avalanches have already claimed nine lives in Washington State, the third-largest annual total for this state. Headlines have been filled with avalanche deaths and injuries in the mountains, including two snowmobilers killed on New Year's Day 2008 when they were swept down several hundred feet on Church Mountain north of Mount Baker, two hikers killed on a hike to Snow Lake near Snoqualmie Pass on December 2, 2007, and a snowshoer who was lost during a hike to Camp Muir on Mount Rainier on December 18,

2007. Avalanches are a constant threat to travel through the major Cascade passes, and avalanche control or unplanned slides often cause substantial delays (figure 8.20). Interestingly, Washington has had far more avalanche fatalities than Oregon (forty-one versus six for 1985–2007), which reflects both meteorological and demographic differences between the states. A great deal has been learned about the nature of avalanches and the associated

8.20. (a) A cloud of snow pushed across Interstate 90 near Snoqualmie Pass due to an avalanche on some nearby slopes. (b) A 17-foot-deep snow slide on US 2 in Tumwater Canyon put an abrupt halt to this tractor trailer's eastward trip across the Cascades. Photos courtesy of the Washington State Department of Transportation.

(a)

(b)

weather, and a modest amount of knowledge can substantially improve your chances of avoiding this hazard.

The essence of the avalanche problem is straightforward: gravity pulls accumulated snow down mountain slopes, while friction or drag at the base of the snow layer and the cohesion of the snow work to keep it in place. Steeper slopes make the problem worse because there is more of a component of gravity down the slope. And the threat is also increased by a smooth underlying surface, a deeper, heavier layer of snow, or a weak layer within the snowpack. In general, trees, large boulders, and other big objects reduce the chance of a slide or limit its extent. A *stable* snowpack is one that is sufficiently strong to resist gravity, while an *unstable* snowpack has one or several weak layers that are prone to fail, particularly when an additional stress is added, such as a rapid load of precipitation, a sudden increase in temperature, or a sudden weighting by a skier, snowmobile, or windblown snow.

Let's examine some of the major elements that control the threat of avalanches, including the preceding weather. Regarding terrain, avalanches generally only occur on slopes steeper than about 25 degrees, with most avalanches occurring on slopes of 30–45 degrees. For a rough comparison, most expert ski slopes do not exceed a 40- to 45-degree angle. Slopes steeper than about 50–60 degrees tend to slough off snow constantly and thus never build up the large mass of snow necessary for a dangerous avalanche. Although a deep layer of snow and a large snowfall rate clearly contribute to avalanche threats, the internal structure of the snowpack is just as important. The central factor is the stability of the snow, its ability to resist the pull of gravity down the slope. Each winter the snowpack develops layering, as snow accumulates from each storm or is affected by periods of warming or rain. Some layers are "strong," comprised of snow grains that are packed closely together and well bonded to each other. Others are weak, typically less dense layers in which the snow crystals are not cohesive: such layers often appear to be feathery or "sugary." The failure of weak layers can allow overlying denser layers to slide down the hill. Snowpacks, particularly in the relatively mild Northwest, generally consolidate and stabilize in time into a solid, strong mass of snow and ice; there is a reason why the white stuff that covers our mountains is often known as Cascade Concrete! Since snowpacks generally stabilize in time, the

8.21. Deep surface hoarfrost, as seen in this example, can lead to a weak layer within the snowpack that can result in subsequent avalanches. Photo courtesy of Dan Zelazo.

greatest avalanche threat is usually when heavy precipitation is falling or on the following day. So a little patience can provide a considerable increase in safety.

One cause of a weak layer is a heavy deposit of surface frost (*hoarfrost*) on a cold, clear night with light winds (figure 8.21). Such frost can sometimes produce a layer of interlocking, feathery ice crystals that forms a less dense, weak layer that is subsequently buried by later snows. Weak layers can also be produced after a snow layer is deposited. A prime example of this is *depth hoar,* large, faceted, cup-shaped crystals that can form in a snowpack with a large change of temperature in the vertical.

8.22. By digging a snow pit one can examine the internal structure of the snowpack, including the existence of weak layers that can lead to future avalanches. In this example, several weak layers are evident in the snowpack. Photo courtesy of Jerry Casson.

8.23. Loose-snow avalanche on Table Mountain in the North Cascades. Photo by Jeremy Allyn.

Snowpacks can fail in other ways. A rapid warm-up to near or above freezing can quickly weaken the bonds within the snowpack, allowing the initiation of a slide. Or a layer of ice or crust can provide a relatively smooth surface upon which an overlying layer slides down a slope. Such a crust layer can be produced by rain falling on snow and subsequently freezing. Rain falling on a snow layer with such an underlying crust is particularly problematic: the rain makes the overlying snow heavier and the water can collect on the crust, providing a very low friction surface upon which the snowpack can slide. Avalanche scientists can evaluate the potential for future avalanches by digging a trench in the snow (also known as a snow pit) and examining the internal structure of the snowpack (figure 8.22). Features such as crusts or weak layers are often quite obvious when viewed this way.

There are two major types of avalanches. *Loose-snow* avalanches begin when loose, unconsolidated snow descends a steep slope. These events generally start at a single point, widen as they

move down the slope, and generally are fairly shallow (figure 8.23). Far more dangerous are *slab* avalanches, in which an entire layer of snow breaks away, often on a preexisting weak layer, leaving a well-defined fracture line at the top (figure 8.24). Slab avalanches can be relatively deep and extensive, making escape more difficult than for loose-snow events. As described below, the greatest avalanche disaster in U.S. history—the 1910 Wellington event—was associated with a half-mile-wide slab avalanche.

Winds can be a major avalanche factor. Sustained winds of 15 miles per hour or more can greatly redistribute loose surface snow, with windward slopes losing snow and slopes facing away from the wind (the leeward slopes) sometimes receiving five to ten times as much as nearby sheltered valleys. Such windblown snow deposition can create unstable cornices on the upper leeward slopes that can fail catastrophically, leading to a major avalanche event (a nice example of such a cornice is shown in figure 8.19). Wind can also break large snow crystals, allowing them to form extremely dense slablike structures that can contribute to an avalanche.

The greatest potential for heavy snowfall and thus avalanches in Northwest mountains is generally after fronts have passed and cool, unstable air floods the region. This showery regime is accompanied by a shift of the low-level winds to the west that enhances upward motion, and thus increases precipitation in the mountains (see chapter 4 for a more detailed explanation). Accompanying frontal passage, the winds often shift from easterly to westerly in the Cascade passes, with a change from light continental snow to warm snow that produces a dense, cohesive layer. Such a snowpack structure, with dense snow over a less dense layer, is particularly weak and often leads to slab avalanches. Another major threat comes from Pineapple Express situations (see chapter 3), in which heavy rain and a rising freezing level can both destabilize and weigh down a preexisting snowpack.

The greatest avalanche disaster in the United States occurred in the tiny Cascade Mountain village of Wellington, Washington, on March 1, 1910. A week earlier, the Spokane local passenger train and a fast mail train had been held in Leavenworth, Washington, because of heavy mountain snows. Allowed to proceed westward on February 23, they made it through the Cascade Tunnel before being stopped by snow and avalanches in the railway town of Wellington beneath the shadow of Windy Mountain. A series of strong Pacific systems continued to move across the region for the next several days, bringing heavy

8.24. A major slab avalanche on Mount Hood. Photo courtesy of Michael G. Halle.

snow and small avalanches. On the worst day, nearly 11 feet of snow fell. On the last day of the month, temperatures warmed, perhaps with a Pineapple Express event, resulting in a change to heavy rain at lower elevations. During the early morning hours of March 1, thunder and lightning storms hit the Cascades, and at roughly 1:45 AM a loud rumble was heard over the slopes above, associated with a half-mile-wide slab avalanche. Charles Roe, a Great Northern Railway employee, described it as "a crescendo of sound that might have been the crashing of ten thousand freight trains."[1] The descending avalanche picked up the trains and pushed them 150 feet into the Tye Valley below (figure 8.25), breaking apart the rail cars and

---

1   Quoted in JoAnn Roe, *Stevens Pass: The Story of Railroading and Recreation in the North Cascades* (Seattle: Mountaineers Books, 1995), 84-90. Much of the material about the disaster is from an article at HistoryLink.org (essay 5127, "Train Disaster at Wellington Kills 96 on March 1, 1910," by Greg Lange, 2007).

8.25. The 1910 Wellington avalanche pushed two trains into the Tye River valley below. This image shows two of the five locomotives. Photo by Asahel Curtis and provided courtesy of the University of Washington Library, Special Collections (negative A. Curtis 17461).

nearby structures. The death toll reached ninety-six and included passengers and train staff, as well as Great Northern personnel from the town of Wellington.

The fact that a huge avalanche occurred on the slopes of Windy Mountain was no accident; not only were the mountain's flanks steep, but logging and forest fires had removed much of the protective vegetation. To deal with the severe threats of avalanches along the route, Great Northern subsequently built a new tunnel through the Cascades that avoided the worst grades and constructed a series of snow sheds at the vulnerable locations on the west side of the Cascades. The town of Wellington was renamed Tye and was abandoned a few years later. Today, interpretative signs on the Iron Goat Trail are the only indications of this great disaster, and recently Gary Krist has written a well-received book on the subject, *The White Cascade: The Great Northern Railway Disaster and America's Deadliest Avalanche.*

Here in the Northwest we are extraordinarily lucky to be served by one of the leading avalanche prediction centers in the world: the Northwest Weather and Avalanche Center (NWAC) in Seattle. Staffed by expert avalanche forecasters, with access to real-time weather and snowpack data throughout the Cascades, the NWAC provides mountain and avalanche predictions throughout the winter season. Their Web site (http://www .nwac.us) provides access to this data and a range of educational materials. In addition, their forecasts and warnings can be accessed over NOAA Weather Radio (described in the weather resources section at the end of this book) and through their public telephone numbers (206-526-6677 for Washington, 503-808-2400 for the Mount Hood area).

9

WEATHER FEATURES
OF THE INLAND NORTHWEST

LACK OF PRECIPITATION IS THE BEST-KNOWN CHARACTERISTIC OF THE WEATHER

east of the Cascade crest, with cold winters and hot summers a frequent elaboration.

Chapter 2 described the typical weather conditions of the inland region, chapter 3 noted the

flash flood and snowmelt flooding in the area, and chapter 4 discussed east-side snow and ice

storms. In addition to these key weather elements, many other weather features of the inland

Northwest are worthy of attention. Wind turbines have proliferated over eastern Washington

and Oregon, driven by a nonagricultural resource of the area: persistent wind. A major cause

of such wind is gaps in the Cascades that act as conduits for air moving to and from the west.

Wind and dry conditions also produce dust storms that can reduce visibility to a matter of feet and bring traffic to a halt. The eastern portion of the Pacific Northwest, separated from the moderating influence of the Pacific Ocean by mountains, is also home to the extreme temperature records of the entire Northwest region, both maximum and minimum. Finally, with regional winds generally from the west, eastern Washington and Oregon are often downwind of the Cascade volcanoes. When eruptions occur, as in the May 1980 uncapping of Mount Saint Helens, dust and volcanic debris can fill the air to the east, producing profound short-term weather changes.

NORTHWEST WIND POWER AND THE KITTITAS BREEZEWAY

The amount of wind-power generation in the Northwest has increased rapidly, with Washington and Oregon producing 2083 megawatts (MW) of electricity from the wind as of March 2008. The greatest potential for Northwest wind power is found in two areas: along the Pacific coast and within or downstream of gaps in the region's substantial mountain ranges (figure 9.1). The potential for wind power increases with the cube of the wind speed, so doubling speed results in *eight* times more wind energy. Along the Washington and Oregon coasts there are strong and persistent southerly winds during the winter, while northerly winds are often strong along the southern Oregon coast during the summer (chapter 6 describes the summertime coastal northerlies). Strong winds are also frequent over the eastern entrance of the Strait of Juan de Fuca, a location that often experiences powerful southeasterly winds during the winter and moderate westerly winds during summer afternoons and early evenings. Although these

(opposite)

9.1a–b. National Renewable Energy Laboratory. Wind energy estimates range from the highest (blue and red) to intermediate (purple and pink) to low (yellow and tan).

west-side locations have substantial wind-energy potential, environmental and scenic issues, as well as difficulties in anchoring turbines in offshore water, have prevented their consideration as sites for wind farms.

During the summer, wind potential is large in and east of the Columbia Gorge, where westerly winds accelerate toward low pressure in the Columbia Basin caused by the warm temperatures. Even greater wind-energy potential exists on the eastern slopes of the Cascades over central and eastern Kittitas County, close to the city of Ellensburg (figure 9.2). Why are Kittitas County and environs so windy? The answer can be found by perusing a topographic map of Washington State (figure 9.3). As noted in chapter 8, the Washington Cascades are not a uniform barrier. Rather, there are two major elevated sections: the Cascades north of Snoqualmie Pass and the southern Washington Cascades from roughly Mount Rainier to Mount Adams. Between these two sections is a lower portion, frequently referred to as Stampede Gap, named after Stampede Pass in its center. Because the Stampede Gap area is relatively low and wide, air can cross the Cascades there more easily than at any other location north of the narrow Columbia River gorge.

During the late spring and summer, a large high-pressure area builds over the eastern Pacific, while high temperatures east of the Cascades

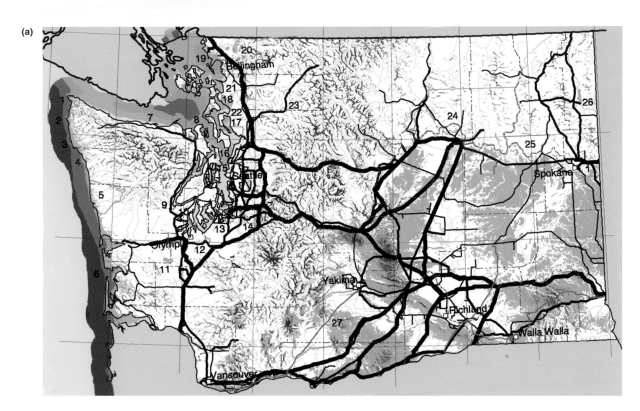

(a)

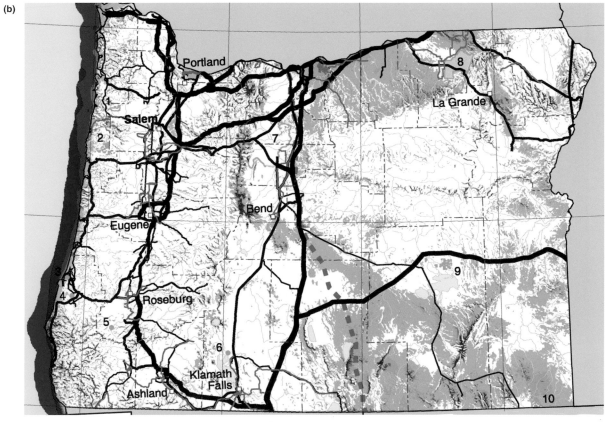

(b)

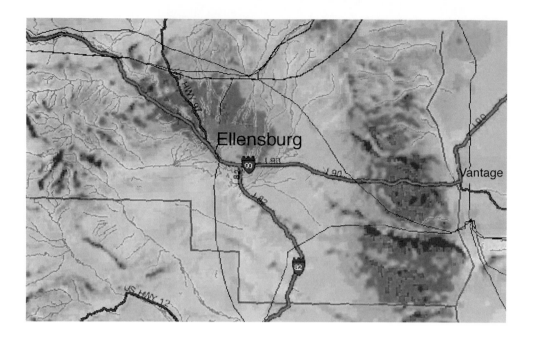

9.2. Wind-energy map for Kittitas County, Washington, and vicinity. Strong winds (purple) are found at lower elevations near Ellensburg and on the ridges to the east. Graphic from the U.S. Department of Energy's National Renewable Energy Laboratory.

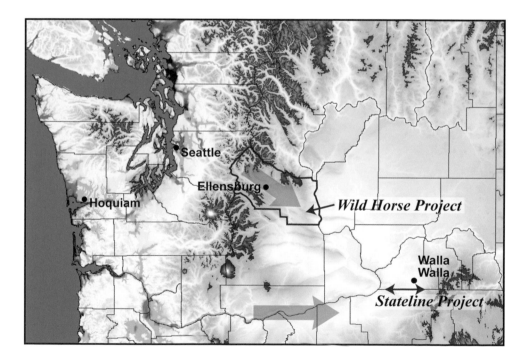

9.3. Topographic map of Washington State, with the border of Kittitas County outlined. Most of this county is east of a lower-elevation section of the Cascade Range. Arrows indicate major wind currents that contribute to wind-power generation in eastern Washington and northern Oregon.

cause pressure to fall, since warm air is less dense and heavy than cold air. The result is a large east-west pressure difference across the Cascades that accelerates air from west of the mountains across the lowest terrain (the Stampede Gap) during the afternoon and evening. This fast-moving current of air then spreads over the region east of the gap: Kittitas County. Strong winds can also occur over the eastern slopes of the Cascades during the winter when Pacific storm systems push powerful westerly winds against and over the Cascades; east of the barrier, the strongest flow often occurs downstream of Stampede Gap over Kittitas County and neighboring areas. Since the lower elevations of the Kittitas Valley are often filled with cold, stable air during the winter, the strong westerly winds sometimes remain aloft, influencing the higher nearby ridges such as Whiskey Dick Mountain.

The unusual wind characteristics of Kittitas County and vicinity are highlighted in figure 9.4, which shows the monthly average wind speeds at various Washington locations. Seattle and Hoquiam have their strongest winds in the winter, with Hoquiam on the Pacific coast having noticeably higher winds, and thus greater wind-power potential, over the entire year. Walla Walla, in the southeast portion of the state and close to the large Stateline wind-energy project, has similar speeds (8–10 miles per hour), but experiences the strongest winds in the spring and summer. In Kittitas County, Ellensburg is a very different animal, with far stronger wind speeds during much of the year, averaging nearly 16 miles per hour in June and July. Since wind energy increases rapidly with wind speed, Ellensburg and the surrounding lowlands are fertile areas for wind-energy generation for two-thirds of the year.

Although winds are light during the winter at Ellensburg, wind surveying by energy companies found that the crests of nearby ridges, above the quiescent cold air near the surface, enjoy substantial westerly winds during the winter as well. Thus, the ridges east of Ellensburg have become prime sites for wind farms. In the Kittitas "breeze-

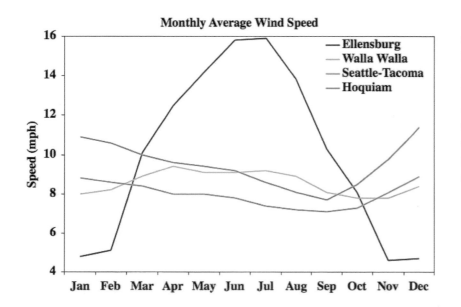

9.4. Monthly average winds (mph) at Ellensburg, Walla Walla, Seattle-Tacoma Airport, and Hoquiam, Washington. The winds are based on hourly observations for 1992-2002. Since the wind-energy potential increases with the cube of the wind speed, the Ellensburg area offers productive wind-power sites.

9.5. The Wild Horse wind turbine project was completed in late 2006 on Whiskey Dick Mountain, a ridge about a dozen miles east of Ellensburg, Washington.

way," the Wild Horse wind project was completed in 2006, located on Whiskey Dick Mountain 13 miles east of Ellensburg (figure 9.5). This project encompasses 127 large wind turbines that have the capacity to produce 230 megawatts of power, enough to supply approximately seventy thousand homes. Other large wind farms are planned for additional locations near Ellensburg.

Another major area of wind-power generation in the interior Northwest is along the Oregon-

Washington border, where strong winds are associated with the only sea-level conduit across the Cascades—the Columbia River Gorge. During the summer, as pressure falls east of the Cascades and builds over the Pacific, westerly flow exiting the Gorge spreads over the border region east of the Cascades, with particularly strong winds over the crests of the area's numerous ridges. To take advantage of these winds, several projects have been built or are in the planning stages as this book goes to press, the largest being the Stateline project, which straddles the state border between Touchet, Washington, and Pendleton, Oregon. Located on the higher ridges, the Stateline wind turbines intercept winds averaging 16–18 miles per hour during the warm season. In 2007 this project had 454 turbines, producing a maximum output of 300 megawatts of electricity.

THE DUST STORMS OF EASTERN WASHINGTON AND OREGON

Each year, residents of the Columbia River basin of eastern Washington and northeastern Oregon experience dust storms that can drive visibility down to zero. Both a health hazard and a danger to the motoring public, these dust storms occur when strong winds lift dust and small particles from agricultural fields, rangeland, and unpaved roads into the atmosphere. East of the Cascades, dust storms have always been a frequent visitor during the dry season from late spring through early fall: the diaries of Lewis and Clark noted such events in October 1805. The potential for dust storms increased as farming spread across eastern Washington and Oregon, since tilled soil, denuded of protective vegetation, is vulnerable to the effects

9.6. Dust cloud approaching Moses Lake, Washington, on August 12, 2005. Photo courtesy of David Dorman.

of the strong winds that frequent this region. Overgrazing has also been a factor. Although dust storms are more likely to occur during the warmer and drier months from May through October, they can occur any month of the year. Specifically, a rainless spell of two to three weeks can dry the topsoil sufficiently to allow blowing dust if the winds are strong.

As a rule of thumb, dust storms require sustained (averaged over two minutes) winds exceeding roughly 20–30 miles per hour. Winds of such magnitude or greater can accompany strong cold fronts pushing out of the north, intense thunderstorms that produce powerful outflow winds that can reach 40–60 miles per hour, or strong westerly flow that pushes eastward across the Cascades. The approach of a dust storm can be dramatic, with the appearance of a dark, turbulent cloud on

the horizon whose passage alters the scene from clear skies to near zero visibility within a few seconds (figure 9.6).

One of the most acute dangers of dust storms is roadway collisions, with many dust events contributing to multicar accidents that have caused injury and death. For example, blowing dust contributed to an eight-vehicle pileup on Oregon State Route 11 northeast of Pendleton, with five injuries (September 25, 2001); to a thirty-eight-vehicle crash near Prosser on Washington State Route 221 that left twenty people hospitalized (October 28, 2003); and to a five-vehicle accident on Washington State Route 241 near Sunnyside (March 16, 2005). In

9.7. Washington State Department of Transportation Web cam imagery at Ritzville (a) before and (b) after the leading edge of a dust storm passed through the area on August 12, 2005.

another incident, five tractor trailers and nine cars collided on September 8, 1990, during a dust storm along State Route 127 in the Washington Palouse between the towns of Dodge and Dusty, the latter name suggesting the frequent occurrence of dust storms.

A more recent eastern Washington dust storm event occurred on August 12, 2005 (see figure 9.6). On that day, an unusually strong summertime cold front pushed southward into eastern Washington. The combination of the cold front and associated strong thunderstorms produced powerful winds,

reaching 50 miles per hour, which spread across much of the eastern portion of the state. The winds lifted enormous amounts of dust that degraded visibility to near zero at many locations. The profound effects of this dust storm are illustrated in figure 9.7, which shows Washington State Department of Transportation Web cam images at Ritzville, Washington, immediately before and after passage of the front. A twelve-vehicle accident associated with the blowing dust temporarily shut down Interstate 90 between Ritzville and Moses Lake.

Probably the greatest Northwest dust storm of the twentieth century occurred during April 21–23, 1931. The preceding spring was unusually wet, with heavy rainfall through the end of March over both sides of the Cascades. In mid-April the situation reversed: unusually warm and dry conditions developed, a region of intense high pressure moved into southwest Canada, and low pressure built over California and Nevada (figure 9.8). As a result, an extraordinary large pressure gradient developed by April 21 over Washington and Oregon, which produced unusually strong northerly and northeasterly winds that reached 50–60 miles per hour over both sides of the Cascades.

The combination of recently tilled fields east of the Cascades, sufficient time for the soils to dry, and unusually powerful winds initiated a combined wind and dust storm that produced the Northwest version of the High Plains dust bowl. Dust clouds spread over the Columbia Basin, turning the sun into an obscured reddish orb; car and house lights were required during daytime. The dust was raised to a height of over 10,000 feet, as observed by an airplane flying from Pasco to Portland that lost track of the ground and overshot its destination, reaching the Oregon coast before it could regain its bearings. The fine dust pushed westward through the Columbia Gorge as a copper-colored cloud, while at the same time surmounting the lower passes of the Oregon Cascades.

Dust spread into western Oregon as far south as Roseburg, greatly reducing visibility and bringing an infernal darkness, and reached northward into southwestern Washington as far as Tacoma. Driven

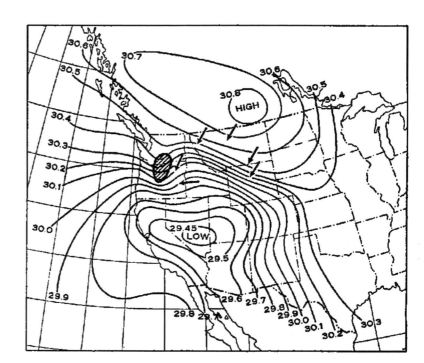

9.8. Sea-level pressure analysis (in inches of mercury) for 5:00 PM on April 21, 1931. High pressure over southwestern Canada and low pressure over the southwestern United States produced large pressure differences and strong winds over the Northwest. The shaded region indicates the area of dense dust at this time. Graphic from Cameron (1931), courtesy of the American Meteorological Society.

9.9. The strong winds during April 21-22, 1931, downed thousands of trees on both sides of the Cascades. This picture shows the damage at an orchard near Mount Hood, Oregon. Photo courtesy of the National Oceanic and Atmospheric Administration, U.S. Department of Commerce.

by the strong winds, the dust plume moved out into the Pacific, where the freighter *Maui*, on a return trip from Hawaii to San Francisco, ran into the dust cloud 500 miles off the coast. Ships approaching the Oregon coast reported dust as thick as dense fog and came into port enshrouded in grit.

In addition to stirring up large amounts of dust, the strong winds during the April 1931 event felled thousands of trees, which damaged homes and businesses and resulted in power outages on both sides of the Cascades (figure 9.9). The combination of powerful winds, high temperatures, and extremely low relative humidity led to numerous forest and wildfires, adding a pale of smoke to the dust-born obscuration.

TEMPERATURE EXTREMES
EAST OF THE CASCADES

Although the Pacific Northwest climate is generally quite temperate, certain east-side locations experience extraordinary cold and heat, with records reminiscent of the Arctic and the Sahara. Isolation from the moderating influence of the Pacific and the amplification of temperature extremes by local terrain help explain the tendency for Northwest temperature records to occur east of the Cascade crest.

For Washington, the coldest temperatures on record occurred on December 30, 1968, at Mazama and nearby Winthrop, both reaching −48 °F (see figure 9.10 for locations). Both towns are at moderate elevation (roughly 2,000 feet) in the Methow Valley, which extends northwestward into the eastern flank of the north Washington Cascades. These extreme temperatures contrasted with the remainder of eastern Washington, which chilled *only* into the −20s °F. In Oregon, the cold record of −54 °F was set on February 9, 1933, at two valley locations within the eastern Oregon highlands: Seneca (4,700 feet) in the Bear Valley and nearby Ukiah (3,340 feet), about 70 miles to the north. All of these cold records occurred after the entrance of Arctic air into the Northwest. Such Arctic "blasts" generally are associated with a large area of high pressure that steers frigid air southward from northern Canada. Although clouds and snow often accompany the entrance of the cold air, within a day the skies generally clear, the winds lessen, and temperatures plummet as the earth radiates heat into the clear skies above.

What makes these locations so cold? Consider Mazama and Winthrop, within the wintertime "icebox" of Washington State. Both are found in the narrow Methow Valley at elevations ranging from 1,760 feet (Winthrop) to 2,170 feet (Mazama) and are surrounded by substantially higher terrain (figure 9.11). Cold air, which is heavier than warm air, tends to descend from the higher terrain and settle into the valley (figure 9.12), particularly during periods of high pressure when the sky is relatively clear, since clear skies allow effective infrared radiational cooling by the ground to space. The fact that the Methow Valley is several thousand feet above sea level often places it above the fog and low clouds that frequently fill the Columbia Basin in winter. Such clouds act as a blanket that lessens radiational cooling to space— thus, the lower elevations of the Columbia Basin are usually warmer than Mazama and Winthrop. The elevation of the Methow Valley also results in more frequent snow coverage during the winter than in the lower Columbia Basin, with snow

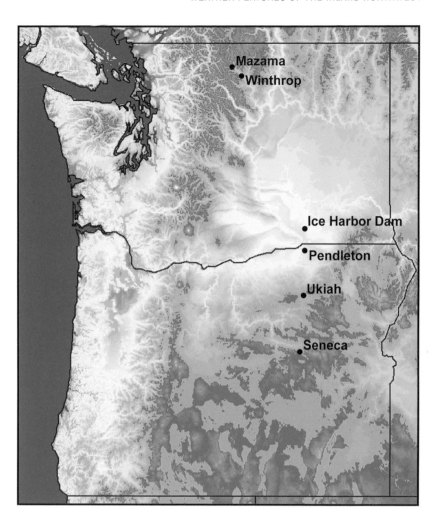

9.10. Extreme temperature locations of the Pacific Northwest. The map shading indicates terrain height, with grey and white representing the highest elevations.

enhancing cooling in two ways. First, snow reflects much of the sun's radiation away from the surface, promoting cooling during the day. The highly reflective nature of snow is obvious when skiing on a sunny day, with sunglasses being a necessity. Second, snow is an extraordinarily good emitter of infrared radiation and thus tends to cool far more effectively than bare soil or vegetation. Finally, both Mazama and Winthrop are east of the Cascade crest and thus are "protected" from the mild, marine influence that moderates temperatures over western Washington. For all of these reasons, the high mountain valleys east of the Cascades, such as the Methow, are often the coldest locations in Washington State.

The Oregon cold spots—Seneca and Ukiah— have much in common with the Methow Valley because they are also located within high mountain valleys or basins east of the Cascade crest (figure 9.13). As with the Methow, radiative cooling to space over the surrounding snow-covered higher terrain cools air near the surface, air that subsequently sinks into valleys and topographic bowls— in the case of Seneca, the bowl-shaped Bear Valley.

During the summer, the most extreme temperatures in the Northwest are found at the lower

(a)

(b)

9.11. (a) The southern end of the Methow Valley on December 24, 2006, viewed from the top of Patterson Mountain near Winthrop and looking toward the northwest. (b) Three-dimensional image of the Methow Valley, viewed from its southern terminus and with the terrain exaggerated by a factor of three. Winthrop is at the valley entrance and Mazama is at its northern end. Graphic generated using Google Earth Pro, printed courtesy of Google Inc.

9.12. Schematic of the origin of cold temperatures in high mountain valleys such as the Methow. With higher snow-covered mountains around such valleys, surface cooling under clear-sky conditions results in cold air sinking along their slopes and pooling at the valley bottoms. In contrast, over the Columbia River basin persistent low clouds greatly reduce infrared cooling to space, as does the lesser snow cover. The numbers are typical surface air temperatures. Illustration by Beth Tully/Tully Graphics.

9.13. Three-dimensional image of Bear Valley, Oregon, facing east and with the terrain exaggerated by a factor of three. Cool air often sinks into the valley, producing record cold at the cooperative observing location north of the town of Seneca. Graphic generated using Google Earth Pro, printed courtesy of Google Inc.

elevations of the Columbia Basin. The maximum temperature records for Washington and Oregon occurred at two basin locations in close proximity: Ice Harbor Dam, Washington (118 °F on August 5, 1961), and Pendleton, Oregon (119 °F on August 10, 1898). Both locations are at relatively low elevations (475 and 1,074 feet, respectively; see figure 9.10).

Why are these particular locations the regional hot spots? Warm temperatures are favored east of the Cascades, since that area is isolated from the cooling effects of the Pacific Ocean and is generally cloud-free for most of the summer. During summer days, surface radiational cooling to space is completely overwhelmed by the intense heating by the sun. The lowest elevations are warmed by compressional heating, since any air that descends into such low spots is compressed and thus warmed (just as air is warmed when compressed in a bicycle pump, making the pump warm to the touch). The arid nature of the Columbia Basin also contributes to warmth, since vegetation and the moisture that evaporates from it have a significant cooling effect.

VOLCANO WEATHER

Although major volcanic eruptions are infrequent, when they do occur the ash and debris can have a major short-term impact on regional weather. Eastern Washington and Oregon are particularly vulnerable to such volcanic effluent, since winds above the Pacific Northwest are often from a westerly direction. A good example is the 1980 eruption of Mount Saint Helens, whose ash cloud turned day to night over eastern Washington, with substantial impacts on temperature for several days.

After months of minor eruptions and earthquakes, a massive landslide on the flanks of Mount Saint Helens at 8:32 AM on May 18, 1980, initiated a large injection of volcanic dust and gases into the atmosphere (figure 9.14). A column of volcanic ash and dust extended 12–15 miles above the surface, and the debris was rapidly transported eastward by strong winds aloft, reaching Idaho by noon and Montana a few hours later. The ash plume was clearly visible in weather satellite imagery (figure 9.15). Less than fifteen minutes after the eruption (8:45 AM), the ash cloud was already evident as a dark, roughly circular feature over southwest Washington. An hour later (9:45 AM), a portion of an oval-shaped ash cloud had crossed the Cascades and had just reached Yakima. Finally, by 10:45 AM, the leading section of an amorphous dust cloud had arrived at the Idaho border after spreading across the southern half of eastern Washington, while a small but higher portion of the plume remained over the mountains. The dust veil continued to move east and south across Idaho and Montana during the remainder of the day.

As the ash cloud moved over the Cascades and eastern Washington, unnatural midday darkness spread with it. Visibility dropped to a few hundred feet at many locations. With a thick veil of dust overhead preventing the warming rays of the sun from reaching the surface, the normal rise of temperature was arrested as soon as the dust cloud arrived. On the other hand, temperatures could hardly fall since the dust cloud acted as a thick blanket, preventing the loss of infrared radiation to space. Thus, in those places influenced by the dust, the daytime maximum temperatures were greatly reduced by the volcanic cloud. In contrast, nighttime temperatures were enhanced since the earth could not cool off.

The effects on both maximum and minimum temperatures are illustrated in figure 9.16, which shows temperatures at locations within and outside the volcanic dust cloud. Unaffected by the volcanic dust, Boise, Idaho, experienced a normal

9.14. The dust and ash plume from the May 18, 1980, explosion of Mount St. Helens reached far into the stratosphere. The debris moved eastward at approximately 60 miles per hour, pushed by strong winds aloft. Photo courtesy of the U.S. Geological Survey.

8:45 AM

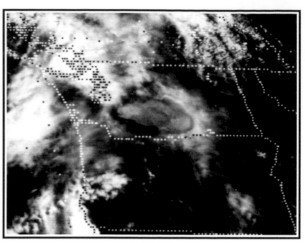

9:45 AM

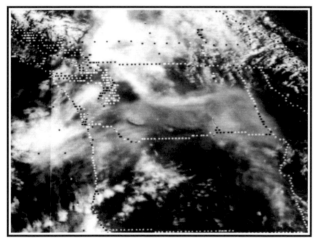

10:45 AM

9.15. A series of visible weather-satellite photos at 8:45 AM, 9:45 AM, and 10:45 AM PDT on May 18, 1980, showing the eastward movement of the ash and dust plume from Mount St. Helens. Images from National Oceanic and Atmospheric Administration/National Weather Service GOES-West weather satellite.

daily temperature cycle, with warming during the day and cooling at night, superimposed on a slow warming trend over the period. Yakima had a normal daily temperature cycle on May 17, but when the dust cloud arrived around 9:00 AM on May 18 the usual temperature rise was halted; then, after a slight cooling the temperature remained nearly constant at 59 °F until midnight. With such a dense dust cloud overhead the sun's heat could not reach the surface and infrared radiation from the surface could not get out, so temperatures did

not vary. As a result, temperatures were lower than normal during the day, but warmer than usual at night. Spokane had a similar story, except that the dust cloud arrived a few hours later. The volcanic debris reached Great Falls, Montana, a little before midnight and rapidly stopped the nighttime cooling on its arrival. For all stations, May 19 brought more normal temperatures, although the daytime highs were somewhat reduced. Two days after the eruption, the dust had thinned enough to allow a normal daily temperature regime to return.

By comparing the previously predicted temperatures for the days of the eruption with observed conditions, it is possible to estimate the temperature impacts of the eruption over the entire region (figure 9.17). At 5:00 PM on May 18, a large area of volcano-produced cooling stretched from Mount

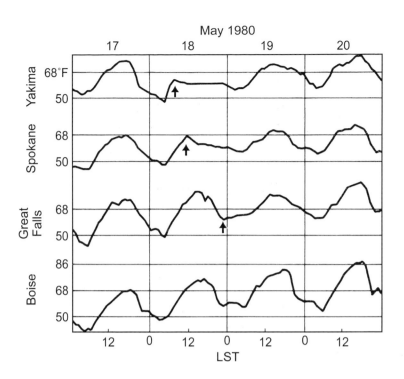

9.16. Surface air temperatures before, during, and after the Mount St. Helens eruption at 8:32 AM (PST, local standard time) on May 18, 1980, at Yakima, Spokane, Great Falls, and Boise. Arrows indicate the time of arrival of the volcanic cloud. Graphic from Mass and Portman (1989), courtesy of American Meteorological Society.

9.17. (a) Estimated temperature changes (°F) at 5:00 PM on May 18, 1980, due to the Mount St. Helens eruption. Shaded areas indicate cooling due to the volcanic ash cloud. Cooling of greater than 14 °F occurred over the southern half of eastern Washington. (b) The same area at 5:00 AM on May 19, 1980. Substantial warming is evident over Montana (greater than 13 °F) and stretches across northern Idaho into eastern Washington. Graphics from Mass and Portman (1989), courtesy of American Meteorological Society.

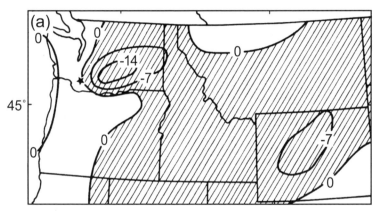

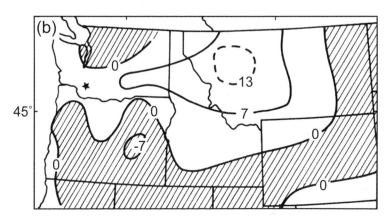

Saint Helens across the southern portion of eastern Washington, with some areas of cooling exceeding 14 °F. In contrast, the volcanic cloud brought warming in eastern Washington and into Montana during the early morning hours the next day (5:00 AM), with the temperatures at some locations enhanced by more than 13 °F.

The ash cloud from Mount Saint Helens influenced more than temperatures east of the Cascades. The static electricity generated by the turbulent volcanic cloud produced extensive lightning displays over the mountain, while a blast of superheated air emanating from the volcano's northern flank descended the mountain slopes, stripping and felling the nearby forests. The rapid melting of huge quantities of the volcano's snow and ice created a surge of water and debris called a lahar, which brought devastation along several rivers.

Ironically, although Mount Saint Helens had a very large local weather impact during the two days after its eruption, it had relatively little influence on global temperatures, unlike major eruptions such as Pinatubo (Philippines, 1991) and Mount Chichon (Mexico, 1982) that caused the whole planet to cool by 1–2 °F. The reason for the small climatic impact is that the Mount Saint Helens eruption contained very little sulfur dioxide gas, which turns into light-reflecting sulfate particles in the stratosphere. Most dust and ash particles ejected by Mount Saint Helens were sufficiently large that they fell to earth within a few days.

Since the winds aloft over the Northwest generally have an eastward component, the ash and dust from Cascade volcanoes will generally move into eastern Washington and Oregon, thus avoiding the major metropolitan areas west of the Cascades. However, there are extended periods of southerly and northerly flow each year in which volcanic dust from major eruptions could greatly affect Seattle, Portland, and other major cities.

10

BLUE HOLES, FLYING FERRIES, AND TORNADOES

||||||||||||||||||||||||||||||||||||||||||||||||||||||||||||||||||||||||||||||||||||||||||||||||||||||||||||||||||||||||||||||||||||||

SOME INTERESTING PACIFIC NORTHWEST WEATHER PHENOMENA ARE EITHER RARE, hard to classify, or are found only in one location. Rain shadows have already been mentioned in chapter 2, but one rain shadow is in a class of its own: the "blue hole" of Sequim, located northeast of the Olympic Mountains, where rainfall amounts typical of Los Angeles are a stone's throw from coastal rain forests. Sometimes the atmosphere plays tricks on our eyes, producing mirages. The Northwest has its share of such apparitions, including flying ferries, dry roads that look wet, and looming hills over water bodies. Finally, tornadoes are infrequent in the Northwest, but do occur on a yearly basis somewhere in the region.

THE BLUE HOLE OF SEQUIM, WASHINGTON

One doesn't have to fly to Los Angeles to enjoy similarly rainless conditions: a trip to Sequim, Washington, will do. Located within the driest area of western Washington, Sequim sits in the Olympic Mountains rain shadow that stretches from the range's northeast slopes toward the San Juan Islands and northern Whidbey Island (see figure 2.6). In this region, air originating from the southwest, the typical wind direction of winter, descends the Olympics, causing drying and warming as it is compressed by increasing pressure at lower levels. In the middle of the most arid portion of the rain shadow, Sequim averages around 16 inches of precipitation a year[1]—less than half that of the Seattle-Tacoma Airport (38.5 inches) and about the same as the Los Angeles Civic Center (14.89 inches).[2] Some years have brought as little as 10 inches to Sequim rain gauges, similar to the typical amounts in the driest parts of eastern Washington. Sequim is so dry that cacti grow in some fields (figure 10.1). Furthermore, it is not only drier, but also sunnier than the remainder of western Washington, with pictures from weather satellites often showing a hole in the clouds above the region. Pilots sometimes note this "blue hole" when they fly over the area, produced when initially cloudy air is warmed and dried as it descends the northeast slopes of the Olympics.

The first white settlers in the Sequim area tried farming, but they found that the inland prairie, called "the desert," lacked sufficient winter rain and turned dry and brown in the summer (figure 10.2). It was not until a group of pioneers built the first irrigation system in the 1890s with a series of flumes from the Dungeness River that the Sequim

10.1. In the semi-arid environment of the Olympic rain shadow, some cacti can thrive. (a) Prickly pear cactus (*Opuntia fragilis*) is native to the rain-shadow region northeast of the Olympics. Photo courtesy of Fred Sharpe. (b) Sequim is one of the few locations in western Washington where cacti are featured plants at nurseries and flower shows. Photo courtesy of Donna Smith.

(a)

(b)

---

1   Annual average precipitation was 15.97 inches at the Sequim 2E official cooperative observing station from 1980 to 2006.

2   The Los Angeles average annual precipitation is from observations at the city's civic center from 1914 to 2006.

10.2. Unirrigated Sequim farmland viewed looking southwest toward the Olympic Mountains.

prairie began to flourish as a farming community. Ultimately, nine main irrigation districts and companies were formed. In contemporary times, the opening of the first ditch is celebrated each May during the annual Sequim Irrigation Festival.

Sequim's origin can be traced to a prosperous S'Klallam Indian village, and one meaning of the town's name in the tribe's language is "quiet waters." This makes sense from a meteorologist's perspective. The Sequim region is often in the center of a small low-pressure area to the northeast of the Olympics, and thus winds are typically rather weak. In contrast, winds can be raging 10–20 miles to the east or west. This situation is analogous to a large rock in a fast-moving stream—downstream of the rock there is often an eddy or swirl where the currents are weak. In this case, the Olympics are the big rock.

TORNADOES

On April 5, 1972, the deadliest and most intense Pacific Northwest tornado on record struck the

Portland metropolitan area. One of only three F3 tornadoes[3] ever observed over Oregon or Washington, with winds between 158 and 206 miles per hour, this storm first touched down along Portland's waterfront and then crossed the Columbia into Vancouver, Washington, leaving a wake of destruction 9 miles long and a quarter-mile wide (figure 10.3). Six people died, three hundred were injured, and damage totals reached over twenty-five million dollars as the tornado swept through a Vancouver grocery store, bowling alley,

10.3. A Waremart grocery store in Vancouver, Washington, lies in rubble after the April 1972 tornado. Photo courtesy of the *Vancouver Columbian*.

shopping mall, and elementary school. The 1972 tornado was embedded in an unusually strong line of intense thunderstorms that crossed the Cascades and produced another F3 tornado later that day outside of Davenport, near Spokane, Washington. The only tornado death tolls in Oregon were associated with nineteenth-century events, both east of the Cascades. One occurred in the small community of Lexington on June 14, 1888, destroying thirty buildings and killing four, while three lost their lives during the Long Creek tornado of June 3, 1894.

Fortunately, most Northwest tornadoes do not traverse heavily populated areas. For example, the

---

3   Tornadoes are categorized according to the Fujita (F) scale, devised by tornado expert Theodore Fujita of the University of Chicago. Most Northwest tornadoes are of F0 (40–72 mph), and F1 (73–112 mph) intensities, with only a handful reaching the F2 (113–157 mph) level. F3 (158–206 mph), F4 (207–260 mph), and F5 (261–318 mph) storms are very rare in the Northwest.

powerful Wallowa tornado struck the northeast corner of Oregon during the late afternoon of June 11, 1968. Crossing a mountainous, uninhabited timbered area, there were few witnesses to this event, which was one of the most destructive Oregon tornadoes on record. Approximately 3,000 acres of prime timber were destroyed or heavily damaged, with roughly 40 million board feet of lumber blown down. Estimated to have achieved at least F2 status (113–157 miles per hour), this tornado left a damage path 8–10 miles long and nearly 2 miles wide.

Northwest tornadoes are both weaker and less common than over most of the country. Of the fifty states, Washington ranks forty-third and Oregon forty-eighth for the frequency of such storms. This lack of tornadoes reflects the low frequency of thunderstorms, and particularly severe thunderstorms, over the Pacific Northwest. As discussed in chapters 2 and 3, Northwest thunderstorms are few and weak for a number of reasons, first and foremost being the general origin of our air from off the cool Pacific Ocean. Such cool low-level air inhibits rapid temperature declines with height, a necessary condition for thunderstorms. In addition, the cool water cannot provide much water vapor to the air, since the amount of water vapor air can hold increases with temperature. Condensation of water vapor into liquid droplets or ice crystals produces heat that fuels such storms. While thunderstorms in the central and eastern United States often attain heights of 40,000 feet or more, with intense precipitation, hail, and impressive lightning displays, Northwest thunderstorms rarely exceed 20,000–25,000 feet and often produce only a single clap of thunder.

Between January 1, 1950, and September 30, 2005, there were 94 tornadoes over Washington State, with most of the property damage and all six deaths coming from the April 1972 storm. Of these Washington tornadoes, 46 were F0, 29 were F1, 12 were F2, 3 were F3, and the remainder were not classified (but probably were F0 or F1). Over the same period, 88 tornadoes struck Oregon, with 64 classified as F0, 19 as F1, 2 as F2, 1 as F3, and the remainder unclassified. To appreciate how much worse the threat of tornadoes is east of the Rockies, the total number of tornadoes for the states of Kansas, Virginia, and New York for the same period was 3,204, 547, and 350, respectively, with several of these storms in the most intense categories (F3–F5).

Northwest tornadoes have occurred on both sides of the Cascades, as illustrated by figure 10.4a, which shows tornado locations and their intensities over Washington for January 1950 through September 1994. A corresponding plot for Oregon, but without the storm intensities and for a longer period (1887–1995) is shown in figure 10.4b. These maps reflect both the occurrence of tornadoes and whether people were there to see them; thus, less populated regions probably had more tornadoes than suggested by the figures. Even with that considered, regions of enhanced tornado threat are suggested near Puget Sound, over eastern Washington, and in the northern Willamette Valley near Portland. As discussed below, the effects of Northwest terrain might explain some of the regional tornado "hot spots."

The time of the year of greatest tornado frequency changes as one crosses the Cascades. West of the Cascades, tornadoes avoid January, February, and July and are most common in the spring and late fall (figure 10.5). Most of these west-side tornadoes occur in the cool, unstable

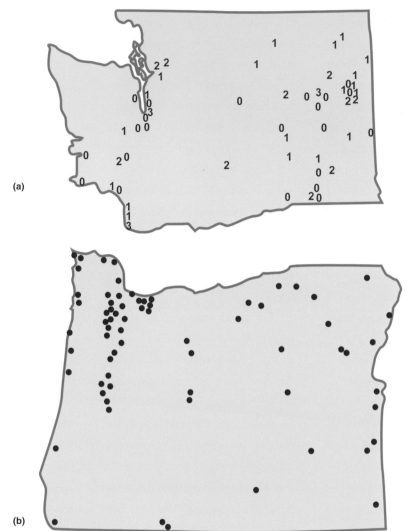

(a)

(b)

10.4. Locations of tornado events over (a) Washington and (b) Oregon. The Washington map is for the period 1950–94 and plots the intensity (Fujita F scale: 0 to 5) at the tornado location. The Oregon map is for the period 1887–1995. Data courtesy of Treste Huse.

air following Pacific fronts, when cold air moves in aloft. With cold air aloft and relatively temperate ocean-warmed air near the surface, the change in temperature with height increases and thus the atmosphere becomes less stable, promoting convection[4] and thunderstorms. Tornadoes usually precede an upper-level disturbance, known as a trough, which typically follows frontal passage; such troughs force upward motion that also encourages convection and thunderstorms.

East of the Cascades, tornadoes are most frequent in April and May, when the atmosphere is most unstable (figure 10.5b). Why is the atmosphere most unstable then? It takes a while for the atmosphere to heat up after being chilled all winter, while the land surface is warmed rapidly by the powerful springtime sun. With cool temperatures

4   The term *convection* signifies clouds that are caused by the vertical instability of the air. Such clouds, which include cumulus and cumulonimbus, often have a characteristic cauliflower shape.

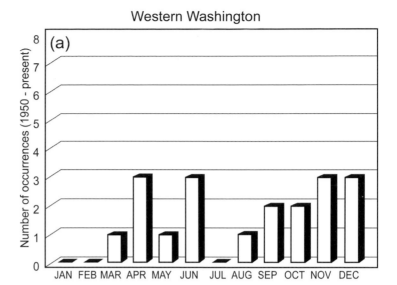

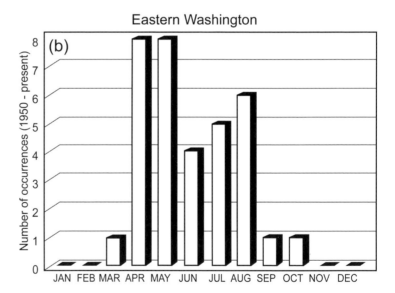

10.5. Numbers of Washington State tornadoes by month, (a) west and (b) east of the Cascades, for the period January 1950–September 1994.

aloft and warm temperatures near the surface, temperatures decrease very rapidly with height, the necessary condition for thunderstorms. Considerable tornado activity continues into the summer over the east side because the lower atmosphere there gets very warm due to a lack of ocean influence. Summertime tornado frequency is also enhanced over eastern Washington and Oregon due to the subtropical moisture that streams northward out of the Gulf of California into the southwestern United States from late June into early September. Such moisture, known as the *Southwest Monsoon*, provides additional "fuel" for the strong convective storms that can lead to tornadoes.

Since thunderstorms and tornadoes are forced by instability that depends on temperature

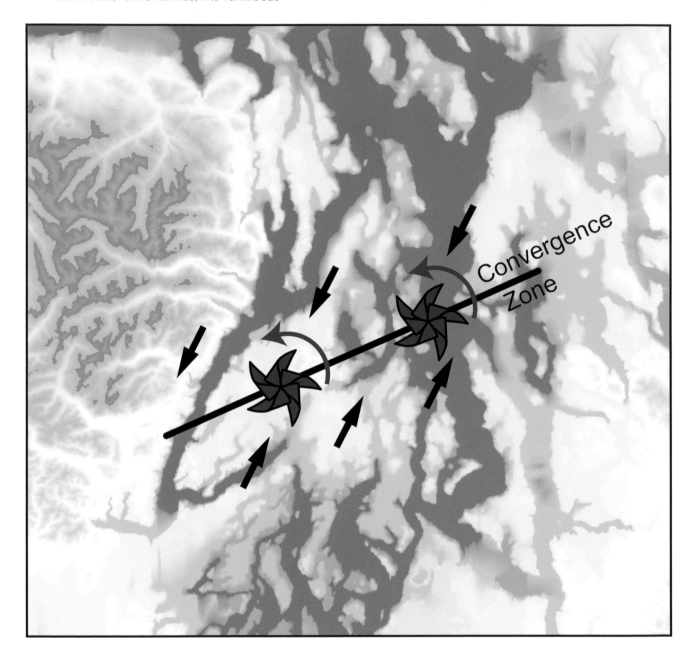

10.6. Schematic of the wind shift across a Puget Sound Convergence Zone. A large pinwheel near the convergence line would tend to spin. This rotation, in the presence of convergence-zone-related upward motion, can produce weak tornadoes. Illustration by Beth Tully/Tully Graphics.

decreasing rapidly with height, it is not surprising that Northwest tornadoes are rare at night and are most frequent from late morning through early evening, when surface temperatures are warmest. West of the Cascades, tornado frequency peaks in the late morning and early afternoon, while in eastern Washington, where surface heating is more important, the peak occurs several hours later and is more sustained.

As noted above, there are hot spots of more frequent tornadoes, such as the Puget Sound basin and the northern Willamette Valley. It appears that the effects of terrain on local winds may explain why these regions are favored. For example, a number of weak tornadoes have developed in the Puget Sound Convergence Zone, a local weather phenomenon that occurs when low-level westerly or northwesterly winds on the coast pass north and south of the Olympic Mountains and converge over Puget Sound (see chapter 7 for a detailed description). The convergence zone frequently forms in the unstable air behind Pacific fronts and can focus convective showers near the convergence line. The large change in winds across the convergence zone, in concert with these convective showers, can cause the air to start spinning, leading to a tornado.

Consider a schematic of the Puget Sound Convergence Zone in figure 10.6. North of the convergence zone the winds are often northeasterly, while to the south the air is coming from the southwest. If a large paddlewheel were placed near the convergence line, it would tend to rotate in a counterclockwise direction. Thus, air near the convergence zone has inherent rotation due to the horizontal change in winds, and clearly rotation is needed for a tornado; however, something is needed to concentrate and enhance the spinning to create a funnel cloud.

In the convective clouds near the convergence line, the upward motion goes from zero near the surface to a higher value up in the cloud (clouds are associated with upward motion). This change

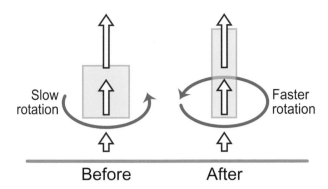

10.7. Increasing upward motion above the surface causes air parcels (indicated in light blue) to be stretched and thinned, resulting in stronger rotation. This is analogous to a skater turning faster when she pulls in her arms while rotating. Illustration by Beth Tully/Tully Graphics.

in vertical motion tends to stretch "parcels" of air in the vertical, with the air parcel becoming thinner as it stretches (figure 10.7). Such a change tends to cause the air to rotate faster, not unlike the more rapid rotation developed by a skater as she brings her outstretched arms to her sides.[5] So, with strong convection and a convergence zone environment with a preexisting tendency to spin, enhanced rotation can develop that can lead to tornadoes.

There have been many examples of convergence zone tornadoes, but a colorful case occurred around 2:30 PM on June 11, 2001. A fairly strong convergence zone had developed over the central Sound, with an active band of cumulus convection at and to the north of the wind-shift line. Suddenly, a funnel developed and started to descend to the surface near West Seattle (figure 10.8 shows an

---

5   The principle underlying this speed-up is called the conservation of angular momentum.

10.8. June 11, 2001, funnel cloud viewed by the KOMO-TV tower cam on Queen Anne Hill. Puget Sound, Elliot Bay, and West Seattle are evident in the more distant part of the photo. The funnel (indicated by the dotted circle) is descending from a cumulonimbus cloud. Photo courtesy of KOMO-TV.

image captured by the KOMO-TV Queen Anne Hill cam). The tornado crossed over the playground of Gatewood Elementary School and several students and teachers were momentarily lifted off the ground in an adventure reminiscent of Peter Pan. Fortunately, no one was injured and damage was minimal.

Other local terrain features also appear to increase the chance of tornadic activity. For example, the Portland metropolitan area is another tornado hotspot and the location of the most dam-

aging tornado in Northwest history: the April 5, 1972, F3 storm described earlier. It is possible that the hills near Portland (the Tualatin Mountains) or the presence of the Columbia Gorge may be contributing factors. Terrain enhancement of tornadoes apparently occurs in other locations as well, such as in the north Cascade foothills near Darrington and Verlot, Washington. Over eastern Washington and Oregon, where surface temperatures are much warmer and instability can be much larger, local terrain enhancement is not as critical for the production of tornadoes.

One tornado might have greatly altered the future of the computer industry and the Northwest economy. Late in the afternoon on September 28, 1962, an F1 tornado (73–112 miles per hour) struck Seattle's View Ridge neighborhood on the northeast side of the city. The tornado funnel first touched ground at the View Ridge playfield, lifting one young football player off the ground. It then knocked down three fences before reaching the home at 7308 44th Avenue Northeast, in which huddled Mary Gates and her six-year-old son, Bill, later of Microsoft fame. With a deafening roar, the carport of the Gates's home was lifted over the roof and sailed into a neighbor's yard. The tornado proceeded to damage five more View Ridge homes, as well as numerous fences and cars, before it struck Hanger 27 at the Sand Point Naval Air Station, ripping off asphalt and tar from a 210-by-400-foot section of the roof. The funnel then traveled across Lake Washington as a waterspout, about 100 feet tall. Reaching the opposite shore, the tornado slammed into Medina, where more destruction occurred. In total, the storm caused property damage of a quarter of a million dollars, with fortunately no loss of life.

FLYING FERRIES AND OTHER NORTHWEST MIRAGES

Although the word *mirage* typically is associated with desert oases, the Pacific Northwest has its share of these magical optical effects. Some of the most dramatic are found near the cool waters of Puget Sound, the Strait of Juan de Fuca, and the Pacific Ocean. Sometimes a stroll along a Northwest shore can reveal strange sights, including ferryboats hovering in the air, with their lower hulls replaced by reflections of their upper structures (figure 10.9). At other times, shorelines appear to loom as vertical walls, many times their actual heights (figure 10.10). On some days, the mountains appear to grow larger in time and then shrink back to normal size; or on a clear, sunny afternoon a wet spot appears ahead on a road, only to disappear as one approaches (figure 10.11). These strange apparitions are produced by light being bent by large changes of temperature within the lowest portion of the atmosphere.

The most frequent type of mirage over western Washington is the *superior mirage*, in which objects appear above their real locations. Such mirages generally occur when a shallow layer of cool air near the surface is beneath a layer of warm air aloft. To understand why this configuration is needed, one must start with the fact that light moves more slowly in cold, dense air than in warm, less dense air. Consider the analogy of a musical

10.9. Two images of Washington State ferries seemingly hovering above the water. Notice how the boats and other objects are elevated, with their upper portions mirrored immediately below. Photos courtesy of Dr. Larry Radke.

(a)

(b)

10.10. Superior mirages taken from the *Victoria Clipper* on June 25, 2006. (a) The shoreline is elevated into a wall by the mirage, with some hint of reflected double images. (b) A cruise ship and the shoreline are magnified, reflected, and inverted, with the Olympic Mountains in the background. Photos courtesy of Dr. Bradley Smull.

10.11. Water-on-the-road mirage. This eastern Washington road appears to be covered with pools of water. The picture was taken around 5:00 PM on July 3, 2006, a warm day with temperatures near 100 °F in much of the lower Columbia Basin. Photo courtesy of Joe Wasson.

band moving together in a line. If the line of musicians begins walking over a grassy field where movement is easy (analogous to the less dense air) and crosses into a mud-covered area (corresponding to denser air), the mud will cause the line to slow down. If the line of musicians approaches the mud at an angle, the natural tendency will be for the line to turn (figure 10.12). Similarly, light will bend when it crosses from a less dense to a denser medium at an angle (figure 10.13). The direction of the turning is such that objects appear to be higher in the sky than they really are (that is why it is called a *superior* mirage). Implicit in the production of mirages is that our brain assumes light is moving in a straight line: we believe that objects are where we see them, which is *not* true for the mirages discussed here.

Superior mirages are frequently observed over Northwest waters during the summer. The water temperatures of Puget Sound, the Straits of Juan de Fuca and Georgia, and the Pacific are quite cool,

typically near 50 °F, while warmer air is found aloft. This large vertical increase in temperature can create superior mirages in which the shoreline turns into cliffs, walls, and turrets while boats and other objects appear to float in the sky (see figures 10.9, 10.10). Sometimes the change in air density with height produces more complicated mirages, with objects being inverted. Superior mirages can make entire mountain ranges rise in the distance, an apparition sometimes observed from Victoria,

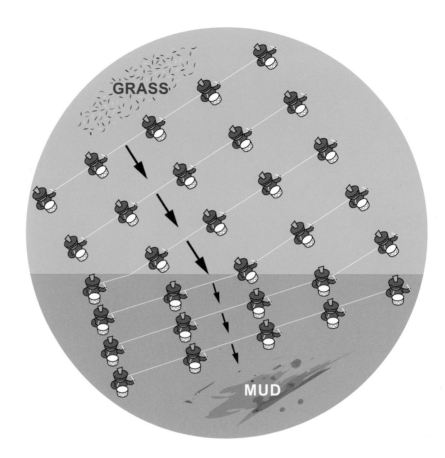

10.12. When a line of musicians marching on grass approach an area of mud at an angle, the slowing effects of the mud cause the line to change direction. Changes of air density can have similar effects on visible light, bending light and producing some mirages. Illustration by Beth Tully/Tully Graphics.

Canada, where the Olympic Mountains appear so large that they seem just on the other side of town instead of miles away across the Strait. Perhaps the most frequent Northwest superior mirage is the setting sun. Because of the general decrease of density with height in the atmosphere and the resulting bending of the sun's light, when the sun appears to our eyes to be slightly above the horizon, it has already set! Similarly, when the sun appears to have just risen in the morning it is actually still below the horizon.

During warm summer days when a shallow layer of air near the surface is strongly heated by hot roads, while much cooler air is a few feet above, the opposite type of mirage occurs. The change in air density associated with these temperature variations (less dense below, more dense above) causes light to bend so that objects appear to be lower than they actually are, thus producing *inferior mirages* (figure 10.14). The classic example is the water-on-the-road mirage, pictured earlier in figure 10.11. A distance ahead, the road appears to be wet, but when you get there, no water is to be seen. Where wet spots appear, you are actually seeing light from the sky or higher nearby objects that has followed a curved trajectory (figure 10.14b). Although water-on-the-road mirages are occasionally observed west of the Cascades, they are often seen over eastern Washington and Oregon during the summer.

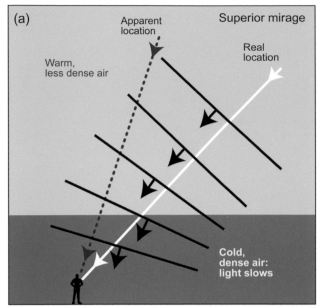

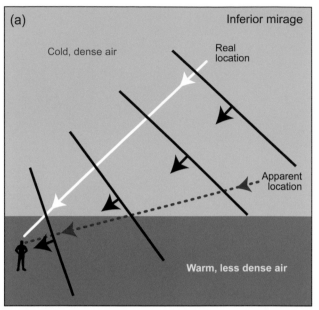

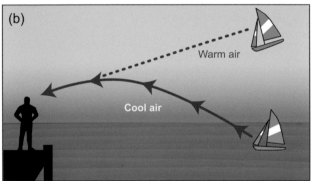

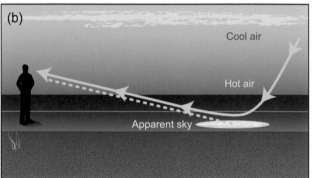

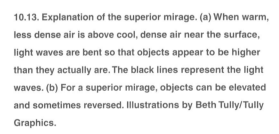

10.13. Explanation of the superior mirage. (a) When warm, less dense air is above cool, dense air near the surface, light waves are bent so that objects appear to be higher than they actually are. The black lines represent the light waves. (b) For a superior mirage, objects can be elevated and sometimes reversed. Illustrations by Beth Tully/Tully Graphics.

10.14. Explanation of the inferior mirage. (a) When cool, dense air is above warm, less dense air, light waves are bent so that objects appear to be lower than they actually are. (b) Looking toward the ground one sees a patch of shiny water, while in reality one is viewing light from the blue sky that has been bent by the large change in density in the vertical. Illustrations by Beth Tully/Tully Graphics.

11

THE CHALLENGE OF PACIFIC NORTHWEST WEATHER PREDICTION

WEATHER PREDICTION IS NOT AN EASY TASK AT ANY LOCATION, BUT PACIFIC

Northwest forecasting offers particular challenges. First, the terrain and complex land-water

contrasts of the region produce a multitude of local weather features as well as large differences

in weather over short distances. For example, winds can blow at 45 miles per hour at the exit of

the Columbia Gorge at Troutdale, Oregon, but be calm at Portland, 10 miles away. Second, the

storm and weather systems entering the Northwest usually originate over the data-poor Pacific

Ocean, since weather generally moves from west to east in the midlatitudes. With fewer observa-

tions over the ocean compared to land, Northwest meteorologists and the computer simulation

models they depend on often operate with an imperfect description of the weather over the Pacific, which has negative implications for the quality of forecasts. This chapter considers these challenges, describes the current technologies used by today's Northwest meteorologists, and explains why weather prediction over the Northwest has improved substantially and should continue to improve during the next few decades. Ironically, the day may come when Northwest weather prediction is the most skillful in the nation.

THE SCIENCE OF WEATHER PREDICTION

It is often said that weather prediction is as much art as science, but this is no longer true. Weather prediction has become a highly quantitative science with a basis in physics and mathematics. Forecasting can be divided into key steps: observing and describing the current state of the atmosphere; running numerical weather-prediction models to provide forecasts of the future atmosphere; statistical postprocessing of the computer forecasts; and interpreting and communicating forecasts to the public and other users.

Observing the Atmosphere

The atmosphere is fully three-dimensional and to understand and predict its future requires observations at both the surface and aloft. Since weather knows no boundaries, to predict the weather at any location depends on good observations around the planet. Thus, even during the cold war, the United States, the Soviet Union, and China exchanged weather information, since it was in no one's interest to embargo weather data.

At the surface, the key observation locations have historically been at major airports, which usu-

ally host a full range of weather-observing sensors (figure 11.1). As of this writing, there are roughly ten thousand airport surface observing stations around the world, but these locations are just the tip of the iceberg. Since the late 1990s there has been an explosion of surface observing sites run by a variety of agencies, schools, and even private individuals, with much of the data available in real time on the Internet. Over the oceans, surface data are collected by a far smaller number of moored and drifting buoys as well as from observations made by commercial ships.

Above the surface, the oldest observing technology is the radiosonde, balloon-launched weather stations that radio their observations back to the ground (figure 11.2). Around the world, about a thousand locations launch radiosondes twice a day, but this approach is mainly limited to land. A number of commercial aircraft take observations in flight, providing plentiful data at typical flight level (30,000–40,000 feet) and as planes fly into and out of airports.

Since the 1950s two new observing systems have revolutionized weather prediction: weather radars and weather satellites. Emitting microwave radiation that can bounce off precipitation, weather radar can measure both the intensity and position of rain and snow (figure 11.3). Modern *Doppler radars* have an additional capability: they can tell the speed of the precipitation toward or away from the radar. Since the movement of rain or snow is mainly controlled by the winds, Doppler radars provide information on air motion as well. Weather radars have proven particularly valuable in pinpointing thunderstorms and have greatly improved the prediction of severe weather such as tornadoes and hurricanes. Pacific Northwest

(a)

11.1 (a) Surface observations are taken at airports around the world using sensors such as the National Weather Service Automated Surface Observing System (ASOS). This picture shows the ASOS instruments at Seattle-Tacoma Airport. The three-pronged device at the top of the tall pole is the wind sensor (an acoustic anemometer), while temperature and humidity are measured by the white, mushroom-shaped unit near the base of the pole. Photo courtesy of Michelle Moshner of the Port of Seattle. (b) Over the oceans, both drifting and moored buoys provide critical weather information. Photo courtesy of the National Oceanic and Atmospheric Administration.

(b)

11.2. In a radiosonde system, a package of weather sensors is tied to a parachute and a helium balloon. The instruments rise to as high as 100,000 feet before the balloon bursts. Photo courtesy of Dr. Mark Stoelinga.

11.3. (a) National Weather Service Doppler radars are located at about 120 sites around the United States. Photo courtesy of the National Oceanographic and Atmospheric Administration. (b) Weather-radar image from the National Weather Service Camano Island radar on January 9, 2006, at 1:58 AM PST. Heavy rain (green and yellow colors represent the most intense) extends over western Washington, with the exception of the rain shadow northeast of the Olympic Mountains, and where the radar signal is blocked by the Olympics.

weather radars can delineate the enhanced rainfall on the windward side of regional mountains, as well as the rain shadows in their lee, and provide information on key features such as the Puget Sound Convergence Zone and Pacific fronts (see figure 11.3b). A particular challenge for Northwest weather radars is the blockage of radar beams by the region's mountains and hills, a problem discussed later in the chapter.

(a)

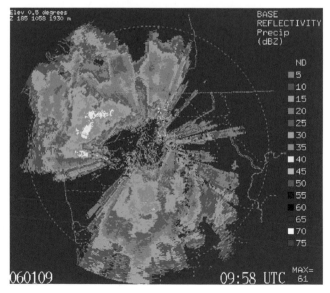

(b)

Weather satellites have proven to be even more important aids to weather prediction. Beginning with the launch of TIROS-1 in 1960, weather satellites have provided a detailed view of both the clouds and surface of the entire planet. Before weather satellites, the world's oceans were great data voids where storms could form and later surprise coastal communities. Intense storms, such as the 1921 Great Olympic Blowdown cyclone, could strike the Northwest coast with little warning; and tropical cyclones, like the great 1938 hurricane, could hit the eastern United States unexpectedly. Today's weather satellites not only track weather systems worldwide, but also possess sensors that can determine three-dimensional temperature and humidity structures from space (figure 11.4). By tracking the movement of clouds from space, winds at several levels can often be determined,

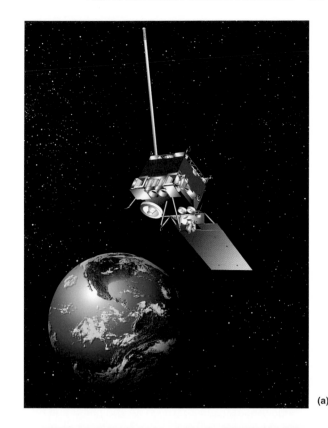

(a)

11.4. (a) Weather satellites stationed over the equator provide continuous imagery of the entire visible hemisphere. (b) A visible satellite image showing the eastern Pacific Ocean and western North America. Taken on November 11, 2007, at 2:00 PM, this image shows a weather system approaching the Pacific Northwest and tropical convective clouds near the equator. Graphic and image courtesy of the National Oceanic and Atmospheric Administration.

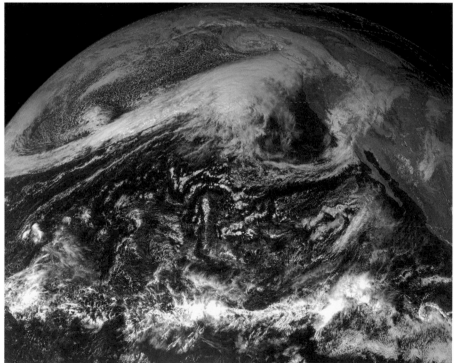

(b)

while other sensors can measure wind speed over the oceans by observing the amplitude of tiny surface waves on the ocean surface.

All of the above observations are distributed to weather-forecasting services around the world, where they are quality controlled to remove bad data. Then, in a process called *data assimilation*, the observations are combined with a previous short-term forecast to provide a complete three-dimensional description of the atmosphere. This short-term computer forecast, called the *first guess*, is used heavily where observations are lacking or sparse, since a short-term prediction of six to twelve hours should be highly realistic. The end product of data assimilation is usually a three-dimensional grid of important atmospheric parameters, such as temperature, wind, and humidity, that encompass the entire atmosphere from the surface to well into the stratosphere. The more grid points and the closer they are together, the higher the *resolution* of the atmospheric description. This description is the requirement for the next step: using computers to predict the future state of the atmosphere, including clouds, wind, temperature, and precipitation.

Numerical Weather Prediction

Before powerful digital computers became available, human forecasters would examine weather maps and observations describing the current and past state of the atmosphere and then predict future weather based on a combination of physical understanding, experience, and extrapolation in time. The predictive skill of such subjective approaches was modest, but useful. The situation changed completely in the early 1950s with the advent of digital computers. Such computers made

possible a new approach to forecasting: using the physical laws that describe the atmosphere to simulate numerically its future evolution, a process known as *numerical weather prediction.*

As noted above, numerical weather prediction begins with collecting observations and data assimilation, which provides a description of important atmospheric parameters on a three-dimensional grid. After data assimilation, the next step in the forecast process is to solve the equations that describe the atmosphere at each of these grid points. But how can an equation predict the future? Let's consider a simple example: Newton's Second Law of Motion.

Newton's Second Law, one of the basic principles of physics and of atmospheric prediction, can be expressed as force equals mass times acceleration: $F = ma$. Force (F) is a push or pull on an object, mass (m) is the amount of matter in an object, and acceleration (a) is the change in the object's velocity with time. In the atmosphere, forces include gravity, differences in air pressure between one location and another, or the friction and drag caused by the rough surface of the planet. Mass depends on the amount of air, the density of which varies with temperature and elevation. Using observations, the forces and mass at each grid point can be calculated, and with simple algebra one can solve for acceleration. That tells us the future because acceleration is the change in wind velocity with time, and observations tell us what the current velocity is. Thus, Newton's second law is a time machine that allows one to predict the future wind if the current wind and other atmospheric quantities are known. There are similar equations that allow meteorologists to predict all other key meteorological parameters, such as temperature, humidity, and pressure. The computer program that solves the complete collection of weather-prediction equations is known as a forecast *model*. This is, of course, not a physical model like one purchases in a hobby shop, but a simulation model comprised of computer code written in a programming language such as Fortran.

Solving these equations takes a great deal of computer power, and the early computers of the 1950s through 1970s had to limit the number of grid points and thus the resolution of the models. During that early period, the grid points were separated by hundreds of miles, and so the computer models could only describe large-scale weather systems. Human forecasters had to fill in the gap, interpreting the coarse computer forecasts to predict local weather phenomena. For example, in the 1970s, computer models could not resolve key regional features such as the Olympic Mountains. As a result, human forecasters in the Seattle office of the U.S. Weather Bureau (now the National Weather Service, or NWS) would predict a Puget Sound Convergence Zone—and enhanced rain north of Seattle—when the predicted large-scale winds on the coast were from the west or northwest.

By the 1990s, weather-prediction models had improved considerably as understanding of the atmosphere grew and rapidly increasing computer power made it possible to solve the prediction equations at many more grid points, allowing the distance between the points to be reduced. As a result, the resolution and realism of the computer predictions increased so that local weather features, which play such an important role in Northwest weather, could be forecast objectively. To illustrate, consider the following Northwest example.

At 1:00 PM on October 20, 2000, the Puget Sound Convergence Zone had become established over northern Puget Sound, with radar showing heavy showers in an east-west band across the Sound (figure 11.5a). Heavy precipitation was also observed on the windward (western) side of the Cascades. A twenty-one-hour computer precipitation forecast using a spacing of 36 kilometers (about 22 miles) between the grid points (similar to the spacing used by the NWS until the mid- to late 1990s) does not resolve the Olympics and thus fails to predict the convergence zone precipitation band (figure 11.5b). Using 12-kilometer grid spacing (about 7.2 miles; similar to the resolution used today by operational NWS prediction models) produces more realistic upslope precipitation on the windward sides of the Cascades, but only hints at the existence of the convergence zone (figure 11.5c). Finally, using 4-kilometer grid spacing (about 2.4 miles; a resolution currently run daily at the University of Washington and used experimentally by the NWS), the Puget Sound Convergence Zone band is nearly perfectly predicted (figure 11.5d).

Based on several years of high-resolution computer weather forecasts, it appears that the above example is highly representative and that many of the local weather features of the Northwest are amenable to numerical simulation and prediction, including mountain precipitation, terrain-induced windstorms, and winds through major gaps in the mountains, such as the Columbia River gorge. Although Northwest weather is often considered difficult to forecast, in some ways the Northwest has the *most predictable* weather in the nation; computer forecast models seem quite capable of handling the complex array of local weather features produced by the regional terrain and

11.5. (a) Weather-radar image from the National Weather Service Camano Island radar for 1:00 PM on October 20, 2000. This image indicates precipitation rate, with green and yellow denoting the heaviest amounts. (b, c, d) MM5 computer model predictions of precipitation at the same time using 36-, 12-, and 4-kilometer grid spacing, respectively. Precipitation amounts increase from light (blue and green) to heavier (yellow and red). The predictions improve dramatically as the resolution of the computer model increases.

land-water contrasts if there is sufficient model resolution and the large-scale atmosphere is predicted with reasonable accuracy.

Strange as it may seem, mountains make the weather *more* predictable. For example, when moist flow approaches from the southwest, Northwest meteorologists can be relatively sure it will rain on the windward (southwest) side of the Olympics and that it will be relatively dry in the rain shadow to the northeast. In contrast, over the relatively flat central or eastern portions of

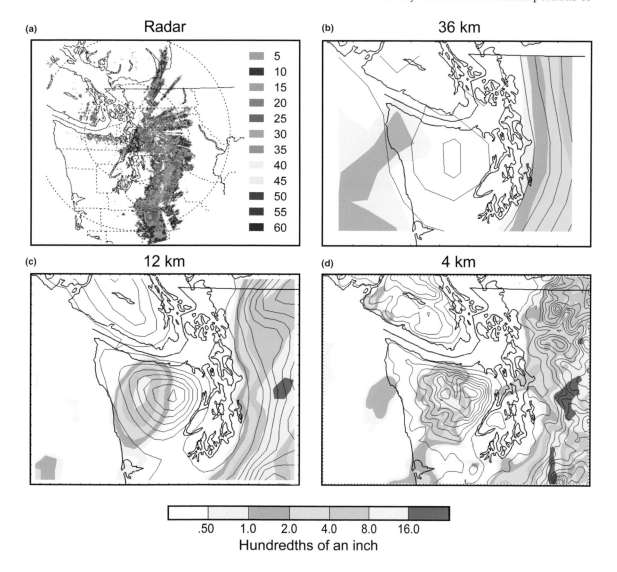

the United States, the effects of local features are greatly reduced, making it far harder to predict precipitation features. Predicting thunderstorms is very difficult, and meteorologists have only marginal skill at such forecasts. Fortunately, unlike the eastern two-thirds of the nation, such storms—and particularly strong thunderstorms—are relatively rare in the Northwest, thus enhancing forecast skill. Northwest meteorologists do have their challenges, as suggested by the occasional forecast failures, with the Achilles' heel of Northwest weather prediction being the lack of observations over the Pacific Ocean. But this weakness may be greatly lessened during the next decade as more comprehensive satellite-based information becomes available (as discussed later in the chapter).

The NWS runs its numerical prediction models at a supercomputer facility outside of Washington, D.C., as part of the National Centers for Environmental Prediction. In the Pacific Northwest a regional group of federal, state, and local agencies and several local universities—known as the Northwest Modeling Consortium—has been running high-resolution weather prediction models optimized for the region for over a decade. These forecasts are available on the Web at Pacific Northwest Environmental Forecasts and Observations (http://www.atmos.washington.edu/mm5rt).

A final step in producing numerical forecasts is *postprocessing*, in which the output from the computer forecast models is improved by statistically correcting the computer predictions. For example, if the computer model forecast at Portland is typically too warm or cold, an adjustment can be made to improve the prediction.

Interpreting and Communicating Weather Predictions

The last step in making a local weather prediction depends on human forecasters. After studying the current and past weather conditions for the area, a forecaster examines the numerical weather predictions that are available (typically forecasters have access to roughly a half-dozen computer forecasts from varying prediction centers). Interpreting these computer predictions in light of experience and physical understanding of the atmosphere, the meteorologist prepares a local weather prediction. Northwest meteorologists must be particularly well trained in the effects of mountains on local weather and skillful in interpreting satellite pictures over the Pacific Ocean.

At the NWS, forecasters prepare their forecasts graphically, indicating regional temperatures, winds, and weather for their areas of responsibility (figure 11.6).[1] This forecast information is then distributed to the media and the public through the Internet and other communication channels using digital graphics and text. In addition, private weather-prediction firms, using the same models and observations, provide custom forecasts to businesses and those who require predictions tailored to specific needs. Local TV weathercasters generally do not make their own forecasts and rarely stray far from NWS predictions.

---

1   The National Weather Service forecast offices for the Northwest include Seattle, Spokane, Portland, Pendleton, Medford, Boise, and Pocatello.

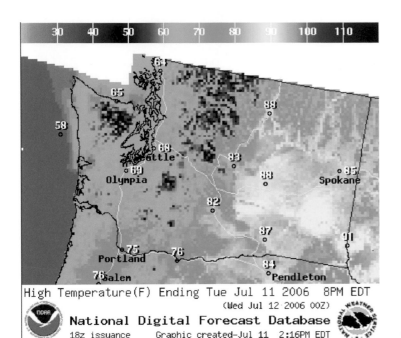

High Temperature(F) Ending Tue Jul 11 2006 8PM EDT
(Wed Jul 12 2006 00Z)
National Digital Forecast Database
18z issuance   Graphic created-Jul 11 2:16PM EDT

11.6. Maximum temperature prediction for July 11, 2006, produced by National Weather Service (NWS) forecasters using a graphical forecast preparation system. Temperatures range from cool (blue) to warm (yellow). On NWS Web sites, you can click on a map to determine the forecast conditions at the selected location.

WHY DO NORTHWEST WEATHER PREDICTIONS SOMETIMES GO WRONG?

Northwest meteorologists are often the targets of humorous comments—I wish I had a dime for every time I have been asked why weathermen never get it right or what kind of dice they use to make forecasts. The truth is that weather prediction *has* improved dramatically as computers, prediction technology, the quality and quantity of observations, and the understanding of weather systems have advanced. Before roughly 1990, major windstorms from off the Pacific were rarely forecast well, and local weather features were poorly understood and rarely skillfully predicted. In contrast, most major windstorms, such as the Inauguration Day Storm of 1993 or the December 2006 Hanukkah Eve blow, are now accurately forecast days in advance; and local weather features, including downslope winds near the Cascades or the onshore push of marine air in the summer, are skillfully predicted.

But even with this considerable progress, it is not uncommon for local forecasts to go wrong, particularly for longer forecast ranges. To understand how this can happen, one first must appreciate that the atmosphere and the numerical models that predict it are complicated *chaotic systems*. Such complex systems can be very sensitive to their initial conditions: a small change in the initial description of the atmosphere can cause significant changes in the forecast that get larger in time. This sensitivity to initial conditions is popularly known as the *butterfly effect*, which suggests that small changes to the atmosphere by a butterfly's flapping wings could eventually lead to major changes in the weather over the entire globe.

Since weather forecasts are sensitive to the initial description of the atmosphere, and errors in describing current conditions are inevitable (because of errors in observing sensors, lack of observations over many areas, and other reasons), there are always errors in the initial conditions that will amplify in time as the computer model is run. Inadequacies in the forecast model itself, perhaps from a poor understanding of the underlying atmospheric processes or lack of sufficient resolution to properly describe key atmospheric processes, can also undermine the accuracy of computer weather forecasts. Typically, forecast errors are relatively small during the first day or two, but by three to five days into the forecast the predictions inevitably deteriorate significantly. For that reason, one should never take the long-range forecasts available on television weather reports or elsewhere too literally; a careful verification of the five-day and greater predictions generally shows little skill over simply forecasting the average conditions for that time of year. Sometimes the atmosphere is particularly sensitive and errors grow unusually rapidly. Such situations are often associated with major forecast failures.

Forecasting accuracy for major weather features such as storms and fronts tends to be less over the western United States than for the eastern United States due to the relative lack of upstream data over the Pacific Ocean. While the eastern United States has a continent's worth of balloon-launched radiosondes, surface observations, and aircraft reports upstream of their area, the Pacific is upstream of the West Coast, with mainly satellite observations and a sprinkling of buoy and ship reports at the surface. Often the lack of data over the Pacific produces minor errors in the forecasts, such as rain coming in a few hours earlier than expected. But occasionally this lack of information is so serious, or the sensitivity of the atmosphere to

a poor start so great, that a major prediction failure occurs. Chapter 5 described a dramatic example: the South Valley Surprise storm of February 2002, in which unpredicted hurricane-force winds devastated the southern Willamette Valley. The NWS forecast models completely failed to predict this storm, probably because of insufficient data over the Pacific.

Another major forecast bust occurred on March 3, 1999. Two days before, the main high-resolution NWS forecast model predicted that a low center would pass across Oregon, drawing cold air from British Columbia across western Washington. As a result, the NWS forecast significant snowfall from Portland to the Canadian border. Unfortunately, the computer prediction was grossly in error, with the low center actually moving northward across Vancouver Island, resulting in a strong windstorm over western Oregon and Washington (figure 11.7). The origin of this failure was apparently a lack of data over the Pacific and the rejection by the quality-control system of a single buoy observation that suggested the correct path.

Even if longer-term numerical weather predictions over the Northwest are problematic, one might expect that short-term forecasts a few hours ahead would be fairly accurate, since one should be able to observe weather systems approaching the coast. Unfortunately, a major weakness in the Northwest weather-observing system often blinds local meteorologists: a lack of coastal weather radars. The Pacific Northwest has the worst coastal weather radar coverage in the contiguous United States, with virtually no coverage along the coast and over the offshore waters. Figure 11.8a shows the situation, with red areas indicating no radar coverage below 25,000 feet. Figure 11.8b suggests

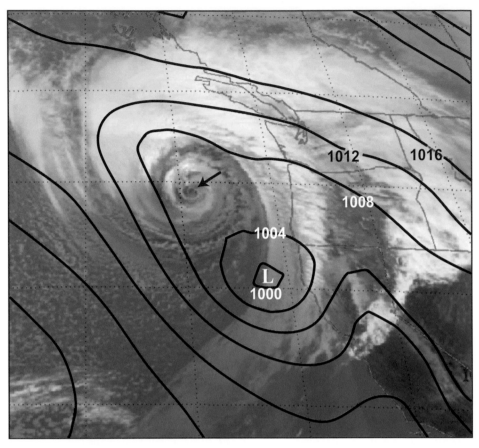

11.7 Weather-satellite image for 4:00 PM on March 2, 1999, and the forty-eight-hour sea-level pressure forecast from the National Weather Service Eta weather prediction model valid at the same time. The swirl of clouds off the northern Oregon coast shows the actual position of the low center and is indicated by the blue arrow. The predicted low center, indicated by an "L," is nearing the northern California coastline.

(a) 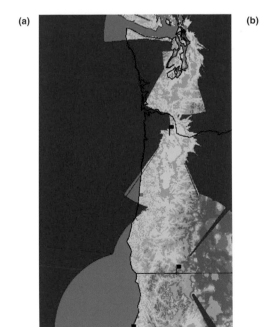 (b)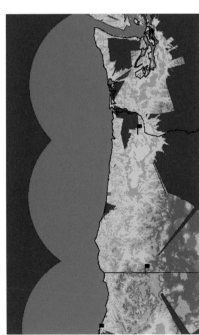

11.8. (a) Radar coverage maps for the current network and (b) a proposed network with two additional coastal radars. Red areas indicate no coverage below 25,000 feet. Graphics from Westrick, Mass, and Colle (1999), courtesy of the American Meteorological Society.

the solution: positioning two radars on the coast, one near Westport, Washington, and the other near Newport, Oregon, would give Northwest forecasters a clear view of incoming weather systems during the six to twelve hours before they make landfall. Although weather-satellite information is invaluable, it provides only minimal information inside clouds and storms; in contrast, weather radar can view the precipitation and winds deep inside weather systems.

LOCAL PREDICTION CHALLENGES

One of the major challenges of Pacific Northwest weather prediction is dealing with local weather variations, many of which are associated with terrain or proximity to water. Local weather can vary mile by mile and neighborhood by neighborhood, and such variations must be understood and communicated. This issue can be illustrated by examining three regional weather features: local temperature effects, low-level fog and stratus, and wind variations between land and water.

Local Temperature Effects

Local TV weathercasters and most Internet sites typically provide only one forecast temperature for each city or town (figure 11.9). Can one temperature represent conditions for any Northwest city? Are the temperatures in your neighborhood the same as at your friend's house a few miles away? The answer, of course, is no. The terrain and water bodies of our region cause temperatures to vary widely over even short distances. There are a number of reasons for this, with land-water contrasts

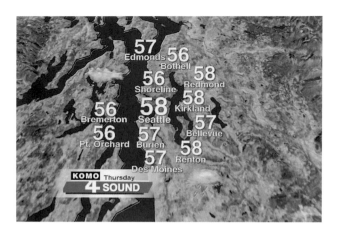

11.9. Local TV stations and Internet sites generally provide one temperature per city or area. When interpreting such numbers, keep in mind that large small-scale temperature variations almost always exist, particularly at night. Photo courtesy of KOMO-TV and Weather Central Inc.

and the drainage of cold air into low spots being two important factors, as discussed in previous chapters.

There are often large differences in temperature between land and water. The temperatures of the region's large water bodies, such as the Pacific Ocean, Puget Sound, and the Strait of Juan de Fuca, do not change substantially during the year, with typical temperatures around 50 °F. In contrast, land temperatures are often warmer than nearby water bodies during the summer, and cooler than the water during the winter. Air temperature, typically measured at about 6 feet above the ground, is greatly influenced by the temperature of the underlying surface and thus takes on the temperature characteristics of the ground or water below.

In the summer, surface air temperatures over the Pacific or connected inland water bodies are typically in the 50s °F, while over land, air temperatures range into the 70s °F and higher. Thus, areas near the water—such as along the coast, the islands in Puget Sound and the Straits of Georgia and Juan de Fuca, or along the shores of Puget Sound—are frequently 5–15 °F cooler than interior locations. During the winter, it is not unusual for western Washington air temperatures over land to drop into the 40s or 30s °F, and occasionally into the 20s °F, which makes land temperatures as much as 20–30 °F less than those over water. In short, coastal zones are often regions of large temperature changes over short distances. Even large rivers such as the Columbia can produce noticeable temperature variations.

Drainage of cold air into valleys and low spots can also produce large temperature differences over short distances. On relatively clear nights the surface radiates infrared energy into space, which cools the surface and the air adjacent to it. Since cold air is denser and thus heavier than warmer air, it tends to settle into valleys and low areas (see chapter 4's discussion of roadway icing). It is not unusual for temperatures to be 2–7 °F less at the bottom than at the top of even modest hills only a few hundred feet high.

Large local temperature changes can also occur due to changes in surface characteristics. A well-known example is the urban heat-island effect, in which cities are warmer than nearby suburbs and rural areas. This warming is caused by a number of factors: concrete and pavement tend to absorb and retain heat from the sun and slowly release it at night; urban areas have large numbers of vehicles, air conditioners, and heating units that warm the outside air; and cities have less vegetation, which cools the atmosphere by transpiration and evaporation of water on leaf surfaces. Portland, Seattle,

and Spokane all have noticeable urban heat-island effects, sometimes making the urban core 5–10 °F warmer than surrounding rural locations.

Finally, air sinking down terrain can cause localized warming. As described several times in this book, when air sinks it is compressed and warmed. Thus, air moving down a major terrain feature can produce localized warming at the base of the barrier. With its substantial hilly terrain, the Northwest has many such subsidence hot spots. During easterly flow across the Cascades, temperatures often peak in the lower western foothills, such as in North Bend, Washington. During southerly flow, the northern lower slopes of the Blue Mountains of northeast Oregon are often particularly warm. Finally, chapter 7 documented the greatest localized warm anomaly of all: the banana belt of the southern Oregon coast near Brookings.

To explore local differences in surface air temperature, five teams of atmospheric sciences students, faculty, and staff drove around Seattle and environs in instrumented cars between 6:00 and 7:30 AM on the cold, clear morning of November 28, 1987. This time was picked because temperature does not change much immediately before sunrise, which occurred at 7:32 AM on that date. The results are shown in figure 11.10. Temperature contrasts were large, ranging from the lower 40s °F over downtown Seattle and near Puget Sound to the mid- to upper 20s °F over the eastern suburbs. Clearly, proximity to water resulted in substantial warming, while interior locations were cooler. Many of the drivers noted that even modest valleys were 2–5 °F cooler than surrounding ridges 50–100 feet above. Such regional temperature variations, reaching 15–20 °F

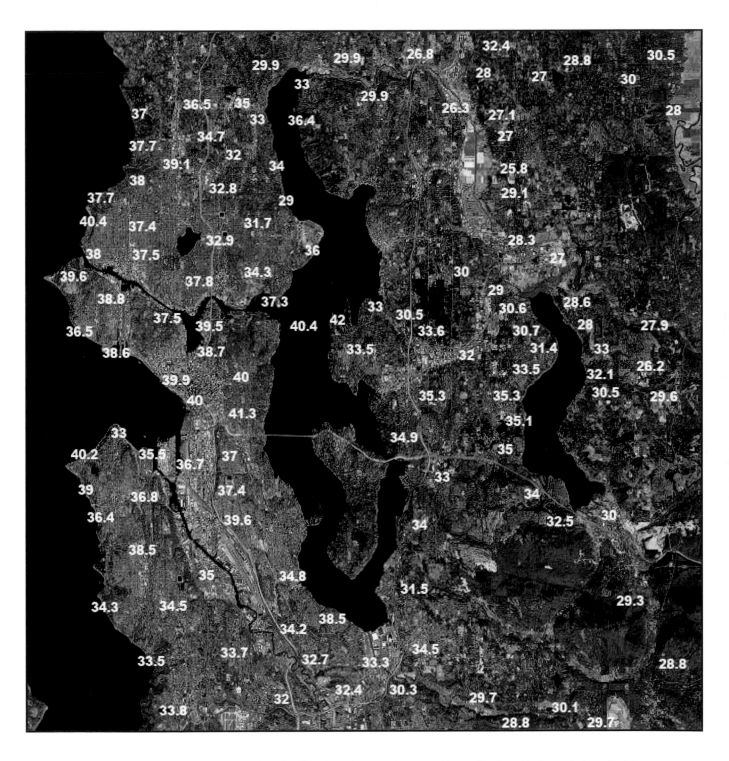

11.10. Air temperatures (°F) at approximately 5 feet above the ground measured by a collection of instrumented cars that drove around the Seattle metropolitan area from 6:00 to 7:30 AM on November 28, 1987. Note the large local variations in temperature, with areas near the water roughly 10–15 °F warmer than over the interior.

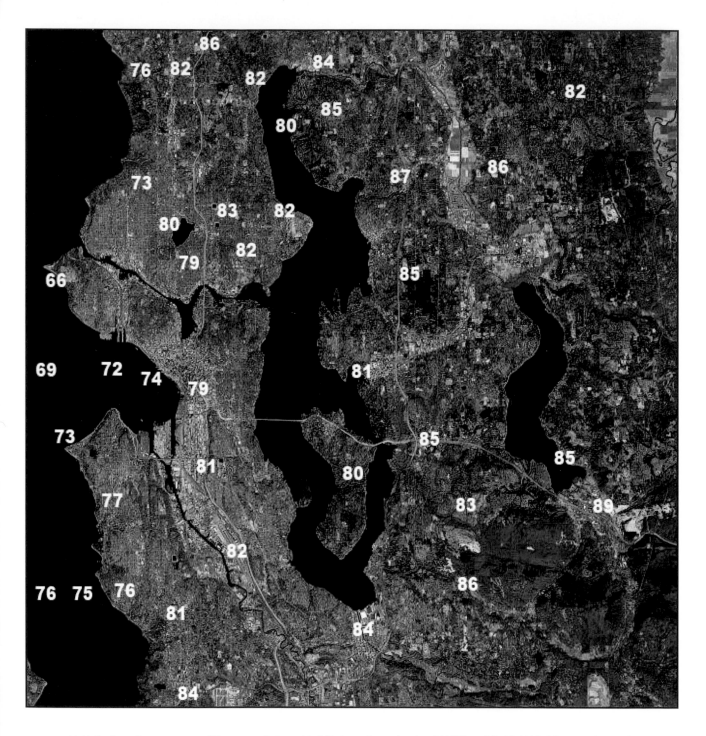

11.11. Surface air temperatures (°F, measured at roughly 6 ft above the surface) at 4:00 PM on July 20, 2006. A large variation of temperature is observed, ranging from the 60s °F over or near the water to the mid- or upper 80s °F inland.

on that morning, are not unusual on cold, clear nights and are often ignored by traditional TV weathercasts and forecasts by other media outlets.

Large temperature variations also occur on warm days as the land heats up and water temperatures remain cool. As a result, those living along Northwest coastlines can nearly always find relief from the heat by heading to the shore. For example, on July 20, 2006, clear skies and warm temperatures were the rule across the Northwest. The surface air temperatures around Seattle at 4:00 PM, shown in figure 11.11, range from the upper 60s °F over Puget Sound and 70s °F along the shoreline to the mid-80s °F away from the water— a 20-degree range. Clearly, a single temperature (such as Seattle-Tacoma Airport's 84 °F) does not represent the range of local values.

As the number of surface observations increase, meteorologists should be able to better document local temperature variations and make such information available online. Furthermore, as computer forecast models are run at higher and higher resolution, the ability to predict local temperature changes should become more skillful. But even without these advances, the knowledge of local temperature variations provided above is useful for many purposes, from selecting plants for the garden to bringing a sweater on a summer trip to a waterfront park.

High Pressure, Inversions, and Fog

One of the most difficult forecasting problems east and west of the Cascades is fog and low clouds. These weather features are critically dependent on the characteristics of the surface, the local terrain, and the detailed structure of the lower atmosphere. Unfortunately, current meteorological models do not handle any of these factors well, because of either lack of sufficient resolution or deficiencies in model physics.

It is ironic that some of the foggiest and most depressing winter or autumn weather in the Pacific Northwest is associated with conditions that would produce the most favorable weather during the warmer portion of the year: high pressure. High-pressure areas and associated clear skies can cool temperatures at the surface, encourage the development of temperature inversions in the lower atmosphere, and lead to fog or low clouds (figure 11.12).

Although fog may not be considered a major societal problem, it is not unusual for airports serving Puget Sound, the Willamette Valley, and

11.12. A foggy Seattle day under wintertime high-pressure conditions. This picture, looking eastward toward Lake Washington, was taken around 2:00 PM on November 20, 2005. The base of the clouds was about 100-300 feet above lake level.

eastern Washington to be seriously affected during foggy periods, greatly reducing their capacity for takeoffs and landings. Dense fog under high-pressure conditions is also associated with dangerous black ice on roadways, and the stable inversion conditions associated with these foggy situations often bring poor air quality and bans on the use of fireplaces and wood-burning stoves. Sometimes fog is extensive and persistent, such as when the basin of eastern Washington fills with fog for days at a time. At other times it can be extremely localized, limited to low valleys or basins, allowing escape with a modest increase in altitude. An extreme example of Northwest fair-weather fog occurred in December 1985 when Seattle-Tacoma Airport experienced the longest continuous run of days with heavy fog on record (thirteen).[2] The fog closed the airport for several days, stranding thousands of holiday travelers, with many planes rerouted to less foggy Portland.

High pressure generally brings favorable weather: a lack of fronts, storms, and precipitation, with relatively few clouds in the middle and upper portions of the atmosphere. The reason for this absence of "weather" is that high-pressure areas are associated with sinking air that evaporates clouds and lessens the threat of precipitation. High-pressure areas also bring warmer than normal temperatures aloft and relatively weak winds near the surface. The clear skies associated with high pressure promote infrared cooling, since the absence of clouds allows radiation emitted from the surface to pass unimpeded to space; in addi-

tion, there are no clouds to send radiation back to the surface. This cooling under clear skies is strongest near the ground, since infrared radiation is more effectively emitted from the surface than from the air above.

During autumn or winter periods of high pressure, the long nights and feeble daytime sun allow the infrared cooling at the surface to dominate over the weak solar warming of the season. Thus, the earth's surface and adjacent atmosphere cool. With dropping temperatures and often a relatively moist lower atmosphere, the relative humidity of the air can increase to 100 percent, resulting in the formation of fog near the ground. Typically such *radiation fog* is shallow, often only a few hundred to a thousand feet deep.

The layer of fog in such circumstances is frequently trapped below an inversion, in which temperature warms with height, the opposite of the normal situation. Inversions are very stable layers in the atmosphere that act as a barrier to the movement of air though them. In addition to limiting the vertical extent of fog, inversions tend to keep pollutants produced by human activities trapped near the surface, contributing to poor air quality. Inversions are forced by two things, either separately or together: cooling near the surface or warming aloft. We have already talked about the cooling, which is typically caused by infrared radiation emitted by the surface. High-pressure areas generally start with warm air aloft, and the sinking air associated with such systems causes further warming aloft that strengthens the inversion. Sinking is greatest several thousand feet above the surface and goes to zero at the ground, since air can't move through the surface. Since sinking air is warmed by compression, stronger downward

---

2   Heavy or dense fog is associated with a horizontal visibility of less than a quarter mile.

11.13. (a) Satellite photo of the Pacific Northwest at approximately 1:00 PM on November 20, 2005. The uniform white areas indicate fog and low clouds. Note how the mountains and the high plateau of eastern Oregon are cloud-free, as are windy locations such as the western side of the Columbia Gorge and the Pacific coast. In contrast, the Columbia Basin, Puget Sound, and the Willamette Valley are enveloped in low clouds. Image from the NASA MODIS AQUA satellite. (b) Sea-level pressure map valid at 10:00 AM on November 20, 2005. High pressure, centered on eastern Washington, dominates the region.

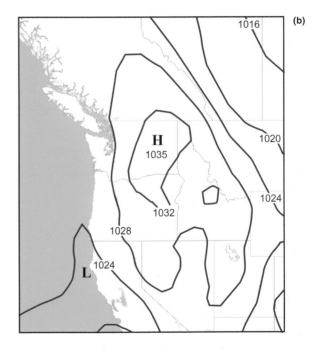

motion, and larger warming aloft strengthens low-level inversions. The weak winds associated with high-pressure areas also assist in keeping inversions strong and fog in place. Wind blowing across the earth's rough surface tends to cause eddies and whirls that mix the air in the vertical. We have all seen such eddies in action on autumn days when the wind moves around or over buildings and causes leaves to swirl around. These turbulent air movements can mix warm air down to the surface and cold air upward, thus weakening low-level inversions and dissipating shallow fog.

When cool-season high pressure builds over the Northwest, low-level cool air, fog, and an inversion often occur over both sides of the Cascades, with Puget Sound, the Willamette Valley, and the basin of eastern Washington being the most usual locations. For nearly an entire week in mid-November 2005, such a situation occurred (figure 11.13). Fog and low clouds filled the lower elevations of eastern Washington, including numerous river valleys and gaps in the terrain. Puget Sound, the Georgia and Juan de Fuca straits, and the Willamette Valley were covered with low clouds, while the Pacific coast and the Portland area were nearly cloud-free. Since the fog and low stratus were only about 1,000 feet thick, the higher elevations of the Cascades and Olympics enjoyed clear skies. Portland and the western Columbia Gorge frequently escape the fog in such situations because the strong easterly winds exiting the Gorge cause sufficient vertical mixing to bring down warmer, drier air from above. The winds also tend to be stronger along the coast, producing mixing that can keep fog and low clouds at bay.

Observations that November day from aircraft and at mountain stations indicated that the fog and low clouds over western Washington had bases between 100 and 300 feet above the surface and tops around 1,200 feet. Figure 11.14 shows the temperature variation with height in the lower few thousand feet over north Seattle. A layer of cool air, with temperatures of approximately 45 °F, was found in the lowest 1,000 feet, above which the temperature warmed rapidly with elevation

(the inversion). By 2,000 feet, temperatures had reached 58 °F. At Paradise Ranger Station (elevation 5,500 feet) and other mountain locations, temperatures reached the mid-60s that day. So one should never judge the potential for a hike by looking out the window.

To explore this intense inversion and the cloud layer below it, I drove to the upper reaches of Cougar Mountain, with a high point of approximately 1,450 feet. Starting the ascent in the fog, I broke out of it at 1,200 feet, where the temperature was in the mid-40s. Ascending 200 more feet, the temperatures felt in the mid-50s, and an opening in the trees revealed a stunning view of the shallow low clouds over the lowlands (figure 11.15). Visibility was exceptional above the low clouds, with Mount Baker, Mount Rainier, and the Olympics distinct and seemingly nearby. This event is a good example of something meteorologists and skiers have long known: on dreary winter days associated with high pressure, it is often possible to escape the fog and low clouds by ascending local peaks or traveling to the foothills of the Cascades (such as around North Bend,

Temperature Change with Height over Northeast Seattle

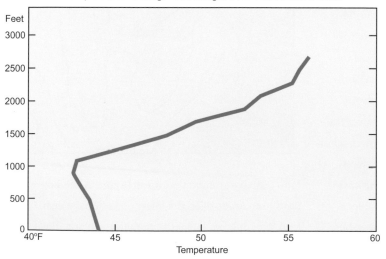

11.14. Temperature variation with height at the National Weather Service forecast office in Seattle at Sand Point at 11:00 AM on November 20, 2005. Note the cool air near the surface, with strong warming (inversion) above. These temperatures were observed by a radar wind profiler that measures temperature by tracking the speed of sound waves at various levels of the atmosphere.

11.15. View near the top of Cougar Mountain at an elevation of approximately 1,350 feet on November 20, 2005. Low clouds and fog filled the lowlands of Puget Sound, while the skies were exceptionally clear at higher elevations.

Washington), where air moving down the western slopes of the Cascades can bring clearing. As noted earlier, the Portland area, the coast, and the highlands of eastern Washington also represent good escape routes. Examining high-resolution visible satellite images that are readily available on the Web can provide useful guidance to the cloud-free areas.[3]

As noted above, autumn or winter high pressure can bring fog that can hamper airport operations. This is less of a problem today than in the past, because many aircraft are equipped with instrument landing aids that enable safe landings with very little visibility. Seattle-Tacoma Airport is particularly plagued by fog, ironically due to its elevation (452 feet above sea level). Fog that forms at night often lifts a few hundred feet during the day and is then considered a stratus cloud.

Airports at low elevations, like Boeing Field, might be fogged in at night, but the daytime lifting of the fog allows airport operations during the day. On the other hand, a higher-elevation airport like Seattle-Tacoma can remain in the clouds even after the stratus or fog layer has lifted elsewhere. During such foggy periods, aircraft are often rerouted to nearby Boeing Field or the generally fog-free Portland Airport, where winds from the Columbia Gorge hamper fog formation. Airports of the Willamette Valley and the Tri-Cities in Washington also have frequent wintertime fog problems, being located in topographic bowls in which cool, foggy air can be highly persistent. Since fog frequently lifts a few hundred feet during the day, it is generally wise to book flights for the late morning or early afternoon during the Northwest fog and low-cloud season.

Although Northwest fog is generally most common during the fall and winter, there are substantial differences in timing and frequency around the region. Figure 11.16 shows the average number of days of dense fog per month for some Washington

3   Infrared satellite pictures, which show the temperature of the clouds or surface, are not useful for seeing fog, since shallow fog is close in temperature to the underlying land or water.

and Oregon locations, based on records from 1971 through 2000. In Seattle, the least foggy period is late spring, with a sharp increase in September and peaking in October. Dense fog declines a bit during the wet midwinter period, followed by the rapid drop in spring. Why is early fall the foggiest period of the year in Seattle? The reason is that in early autumn, nights get much longer, yet the clouds from Pacific storms are not yet a regular fixture. Long nights and relatively cloud-free skies produce good radiational cooling at the surface that can chill the air to saturation and thus initi-

ate radiation fog. Midwinter gets fewer periods of clear skies, so the potential for radiation fog declines. However, the wet midwinter brings fog in another way when rain provides sufficient moisture to saturate the air. Spokane has minimal fog from April through September, but is far foggier than Seattle from November through March. As noted in chapter 9, the basin of eastern Washington often fills with cool, saturated air in winter, resulting in frequent and often persistent low clouds and fog. In contrast, the fog season begins earlier at Quillayute on the north Washington coast,

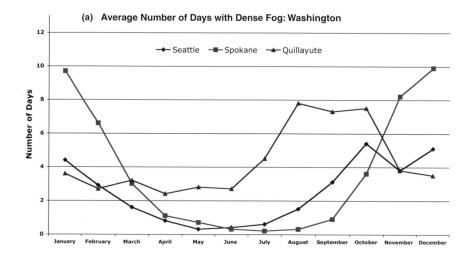

(a) Average Number of Days with Dense Fog: Washington

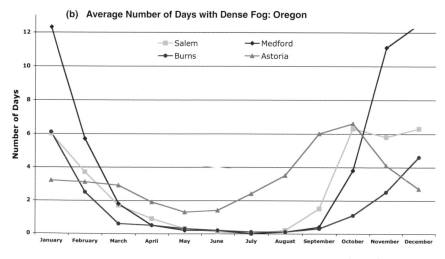

(b) Average Number of Days with Dense Fog: Oregon

11.16. Average number of days per month with dense fog (less than a quarter mile visibility) at selected stations in (a) Washington and (b) Oregon. Based on climatological data from 1971 to 2000.

peaking during the late summer and early fall. Such coastal fog is associated with the extensive low clouds that blanket the eastern Pacific and adjacent coastal regions during the summer and the advent of radiation fog during the lengthening autumn nights. Strong coastal winds during the winter actually work against the development of fog at Quillayute and other coastal locations during that stormy period.

In Oregon, Salem and Portland (not shown), both in the Willamette Valley, have nearly identical fog statistics, with little dense fog from March though September, but around six days per month from October through January. Astoria resembles Quillayute with a well-defined peak during early autumn, and Burns, positioned on the high plateau of eastern Oregon, is nearly fog-free except for November through February. But the "fog king" of the Northwest is Medford, where roughly twelve days of dense fog occur monthly during December and January. The key reason for this plentiful fog is Medford's location, nestled in a relatively narrow bowl within the Siskiyou/Klamath Mountains into which drains the Rogue River. During the winter, many storms pass to the north, reducing cloud cover aloft in southern Oregon and thus allowing good radiational cooling to space. The cool, often saturated air settles into the lower elevations of the bowl, with an inversion overhead that effectively acts as a lid to the cool, foggy brew.

Local Wind Effects

Much of this book has described regional wind variations due to the complex terrain of the Pacific Northwest, with gap winds, diurnal wind changes, and the blocking effects of mountains creating large differences in wind over short distances. But over the western side of the region, another effect is often as large or larger: the variations in wind speed between land and water. Land is aerodynamically rough, with houses, trees, and hills acting to slow down winds near the surface. On the other hand, water offers less resistance to the wind, even with the building of substantial waves. The result is that under the same conditions aloft, wind speeds over water are often 50–300 percent larger than those observed over land. This fact, second nature to sailors and mariners, is often evident to landlubbers when they cross a Columbia River or Lake Washington bridge or take a trip on a Washington State ferry. It is not unusual to experience 5- to 10-mile-per-hour winds over land, while winds over the water are gusting to 20–30 miles per hour. Figure 11.17 provides a nice example of land versus water wind differences, showing wind observations from land stations and Washington State ferries during a period of modest southerly winds. Observing sites away from the water had winds ranging from roughly 3–10 knots (4–12 mph), while over the water winds ranged as high as 28 knots (32 mph). The strongest winds were found over the central Sound, downwind of the longest stretch of water, with a clear strengthening evident on the ferry sensors as the boats left their protective harbors.

Wind strengthens rapidly as air moves from land to water: fast air from aloft quickly mixes down to the surface, particularly when the lower atmosphere is turbulent. Typically within a few miles offshore, the wind accelerates to near its maximum speed over water.

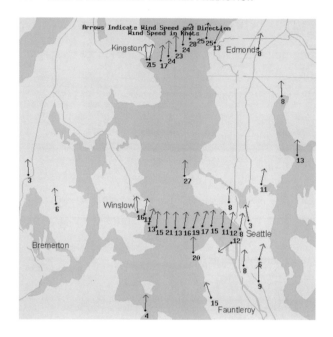

11.17. Wind observations (in knots) at approximately 2:30 PM on December 28, 2007, from surface stations and Washington State ferries. Winds are substantially stronger over the water than over land. Graphic from the Ferry Weather Web site (http://i90.atmos.washington.edu/ferry).

EL NIÑO, LA NIÑA, AND SEASONAL WEATHER PREDICTION

As described earlier in this chapter, the ability to forecast specific weather features is limited to roughly five to seven days into the future, since prediction errors increase in time. But it *is* possible to secure some insight into the general character of the weather several months in advance, and the most potent tool for such seasonal prediction is the connection between the temperature of the tropical Pacific Ocean and the weather around the world.

During the late nineteenth century, scientists realized that there was a regular variation in the surface temperatures of the tropical Pacific. The warm temperature phase became known as *El Niño*

(the little boy), named for the Christ child because this warming usually begins around Christmas-time off the west coast of South America. The opposite phase is *La Niña* and is associated with colder than normal temperatures in the tropical Pacific. Neutral or normal years are humorously called La Nada. The cycle from El Niño to La Niña and back again typically takes three to seven years and is called the *El Niño Southern Oscillation* (ENSO).

By the middle of the twentieth century, researchers put together a more comprehensive picture of ENSO, showing that the variations in tropical ocean temperatures had a large impact on the atmosphere and that the phenomenon represents a complex interaction between the air and water. During normal or neutral conditions, a pool of warm water resides over the western tropical Pacific, with towering cumulus and thunderstorms overhead (since warm water encourages convective activity) (figure 11.18). Winds from the east—the *trade winds*—are found at low levels over the tropical Pacific during the neutral phase and, in fact, help push the warm water to the west. Under El Niño conditions, the easterly trade winds weaken or reverse and the warm water surges to the east, greatly warming the sea-surface temperatures over the eastern Pacific and resulting in an eastward shift of the area of thunderstorms. One can think of the tropical Pacific as a huge bathtub: under normal conditions the easterly trade winds pile up warm water in the western Pacific, while during El Niño periods the trade winds fail and the elevated warm water sloshes eastward. La Niña conditions are just the opposite: the easterly trade winds strengthen, the water surface cools in the eastern and central tropical Pacific, and the warm water and associated thunderstorms shift to the west.

(a) Normal Conditions

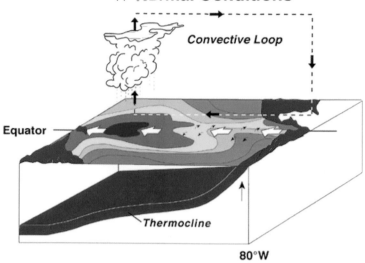

(b) El Niño Conditions

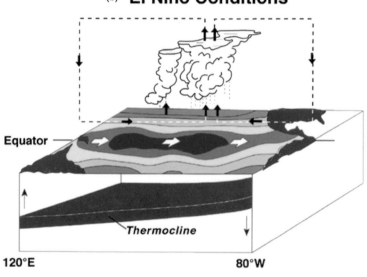

(c) La Niña Conditions

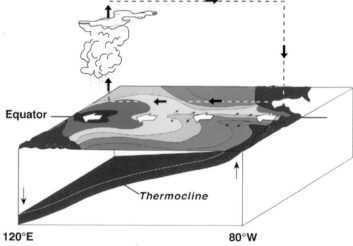

11.18. Schematics of conditions within and over the tropical Pacific during (a) normal, (b) El Niño, and (c) La Niña conditions. Note how warm water shifts eastward for El Niño periods and westward during La Niñas. The thermocline is a layer of large temperature change between the warm surface water and cool water below. Graphics courtesy of the National Oceanic and Atmospheric Administration.

Why should northwesterners care about changing conditions in the tropical Pacific? It turns out that shifts in the tropical atmosphere can greatly influence weather patterns around the globe and that the effect on Northwest weather is considerable. During the 1980s, atmospheric scientists, including several at the University of Washington, described ENSO's global impact and how changes over the tropics were communicated for thousands of miles. Their research found that the collection of thunderstorms over the warmest waters of the tropical Pacific produces atmospheric wavelike disturbances that propagate into the midlatitudes, greatly influencing the weather there. Imagine a large lake. Throwing a large rock into the water causes waves to ripple away from the point of the rock's impact, influencing conditions far away. The atmosphere can be considered as a very large lake and the thunderstorms as large rocks, pushing up into the atmosphere and causing waves that influence weather large distances away. During El Niño and La Niña periods, the location and intensity of the thunderstorms change, and thus the character of the waves and associated weather are altered throughout the world.

As mentioned in chapter 4, during the 1980s the connections between ENSO and Northwest winter weather became increasingly clear. El Niño winters are associated with warmer than normal temperatures and somewhat drier than normal conditions, while La Niña years generally bring wetter and slightly cooler than typical winter weather. These connections between ENSO and Northwest weather vary over the region, as illustrated by figure 11.19. El Niño warming is larger over the northern and western portions of the Northwest, peaking at approximately 2 °F above normal. In contrast, La

(opposite)

11.19. Differences (anomalies) between El Niño and La Niña years and climatology for temperature and precipitation during the winter season (December–February). Graphics courtesy of the Climate Impacts Group, University of Washington.

Niña conditions show more spatial variation, with cooler than normal temperatures over the Cascades and western Washington, but warmer than typical temperatures over a considerable portion of Oregon and eastern Washington.

Concerning precipitation, El Niño winters bring drier than normal conditions over the Cascades, the coastal mountains, and the Rockies, while eastern Washington is slightly wetter than normal. La Niña years are wetter than normal over Northwest mountains. With cooler temperatures and greater precipitation, La Niña years generally bring enhanced winter snowpack to Northwest mountains, resulting in good skiing and adequate water resources for the following summers. In contrast, during El Niño years Northwest residents typically suffer through decreased mountain snows and the potential for subsequent water shortages. Interestingly, although El Niño winters are dry in the Northwest, they often bring heavy rain to the southern West Coast, with landslides on the coastal mountains of southern California.

Although the correlation of ENSO with Northwest weather is useful, it is *far* from perfect. Some El Niño years have had normal or above normal precipitation in the mountains, and not all La Niña years have been wet. There is simply a better than average chance for warm temperatures and low snowpack in El Niño years, like weighting the meteorological dice. Even our ability to predict

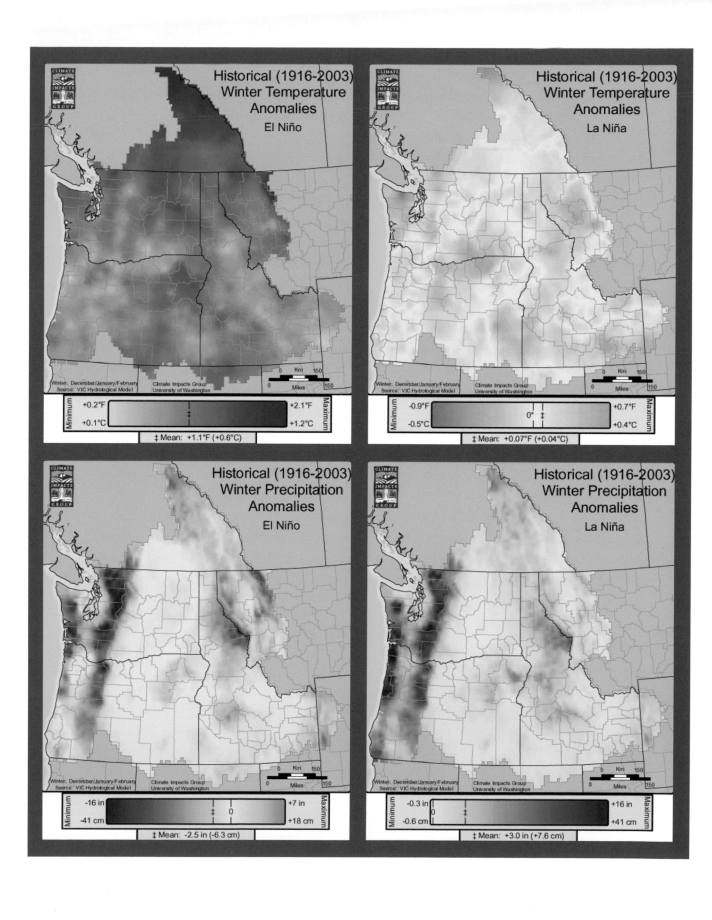

Historical (1916-2003) Winter Temperature Anomalies — El Niño

Winter: December/January/February
Source: VIC Hydrological Model
Climate Impacts Group
University of Washington

Minimum +0.2°F +0.1°C   +2.1°F +1.2°C Maximum
‡ Mean: +1.1°F (+0.6°C)

Historical (1916-2003) Winter Temperature Anomalies — La Niña

Winter: December/January/February
Source: VIC Hydrological Model
Climate Impacts Group
University of Washington

Minimum -0.9°F -0.5°C   0°   +0.7°F +0.4°C Maximum
‡ Mean: +0.07°F (+0.04°C)

Historical (1916-2003) Winter Precipitation Anomalies — El Niño

Winter: December/January/February
Source: VIC Hydrological Model
Climate Impacts Group
University of Washington

Minimum -16 in -41 cm   0   +7 in +18 cm Maximum
‡ Mean: -2.5 in (-6.3 cm)

Historical (1916-2003) Winter Precipitation Anomalies — La Niña

Winter: December/January/February
Source: VIC Hydrological Model
Climate Impacts Group
University of Washington

Minimum -0.3 in -0.6 cm   0   +16 in +41 cm Maximum
‡ Mean: +3.0 in (+7.6 cm)

the phase of ENSO for the next winter is uncertain. Computer models of the tropical atmosphere and ocean have a limited ability to predict the nature of ENSO a few months ahead, but often fail. In recent memory, the official NOAA Climate Prediction Center forecast in August 2006 for the 2006–07 winter was for El Niño to dominate over the entire winter, bringing less than normal precipitation over the region. Instead, the El Niño collapsed unexpectedly in January, early winter precipitation was well above normal, and the seasonal snowfall was near normal. Generally, ENSO stays in the same phase (El Niño, La Niña, neutral) for the whole winter, and this provides at least a modicum of useful guidance for the tenor of the upcoming winter, but there are plenty of exceptions to the "rule."

Many water-resource agencies in the Northwest now use ENSO as a tool in water planning. In 1992, a major El Niño brought reduced snowpack over the Cascades, but unaware of the ENSO-snowpack connection, some utilities, such as Seattle's, dumped water from reservoirs during midwinter as part of their normal flood-control procedures. As a result, there were inadequate water supplies the following summer and mandatory water restrictions were enforced in Seattle and other Northwest cities. By the mid-1990s, understanding of the effects of El Niño was more widespread, so when a major warm event struck in 1997–98, local water agencies were ready, retaining more water in reservoirs during midwinter and using less water for in-house operations such as pipe flushing. As a result, sufficient water supplies were available and no mandatory restrictions were needed during the summer of 1998.

The cycle among La Niña, El Niño, and neu-tral years has a substantial impact on the general weather patterns of the eastern Pacific that explains many of the correlations noted above. El Niño years are often associated with a splitting of the incoming jet stream into two branches: one going into southeastern Alaska and the other into southern California. The latter branch explains the greater tendency for excessive precipitation and storminess over the southwestern United States during El Niño years. With the jet stream splitting, storms approaching the Northwest tend to be sheared apart, and thus the region generally experiences fewer major storms. La Niña years tend to be highly variable, sometimes with a powerful jet stream approaching the Northwest coast, while at other times strong high pressure builds north into Alaska, which can direct cold air southward into the Northwest. Thus, La Niña years typically produce the most snow in the western lowlands. In short, although the connection between ENSO and future Northwest weather is imperfect, it is essentially the only game in town for Northwest meteorologists looking to provide useful information more than a week or two in advance.

ARE THE FORECASTS IN THE OLD FARMER'S ALMANAC ACCURATE?

Grocery stores and newsstands throughout the Northwest sell the latest issue of the *Old Farmer's Almanac*, which contains weather predictions for the upcoming year. Published since 1792, the almanac uses a proprietary method for weather prediction based on solar cycles that supposedly provides 80 percent accuracy for months in advance. Can this possibly be true? Does the tooth fairy exist? The answer to both questions is, unfortunately, no.

Dr. Nick Bond, an experienced weather scientist at the Pacific Marine Environmental Laboratory in Seattle, decided to put the *Farmer's Almanac* to the test by verifying its winter forecasts for thirteen years over the western portion of the Pacific Northwest. The results were not promising. Comparing the almanac's temperature forecasts against the observed conditions showed absolutely no accuracy in predicting whether temperatures would be above or below normal. Flipping a coin, with heads being above normal and tails below normal, would have been just as reliable. For precipitation, the almanac was decidedly worse, being less skillful than a coin flip. In short, the *Farmer's Almanac* provides no useful information regarding next year's weather in the Northwest or anywhere else. However, the homespun advice, wise aphorisms, and astronomical tables might still make it worth the price.

THE FUTURE OF PACIFIC NORTHWEST WEATHER PREDICTION

The skill and value of weather predictions over the Pacific Northwest should improve substantially by 2015–20 for several reasons. First, the "data void" over the Pacific Ocean should be substantially filled by new sensors placed on the next generation of weather satellites. For example, one new instrument, the hyperspectral sounder, will provide thousands of high-quality temperature and humidity soundings[4] over the Pacific each day. Second, faster computers will allow computer forecast mod-

els to run at higher resolution so that important but smaller-scale aspects of Northwest weather, like the Columbia Gorge winds, will be properly handled. In addition, new data-assimilation technologies will better apply current and future observations, and improved physical understanding of weather processes and phenomena should translate into better models. But just as important as these innovations is a new approach to weather prediction that is becoming mature: probabilistic weather prediction.

Since the three-dimensional description of the atmosphere is always imperfect and computer forecast models possess deficiencies, all weather predictions are uncertain and these uncertainties increase in time. Weather predictions today generally ignore these uncertainties; for example, most television stations provide only a single number for tomorrow's temperature as if that were the only possible forecast. It would be far more honest to communicate the inherent uncertainties in weather prediction, providing ranges and probabilities of potential outcomes. Such information would allow the public and other users to make far better decisions regarding weather-related activities. Fortunately, meteorologists have a new weather prediction approach that promises to aid in this new type of prediction: *ensemble forecasting*.

The idea of ensemble prediction is to make many numerical predictions rather than one, with each forecast starting slightly differently or using slightly different computer models (perhaps handling surface friction in different, but equally reasonable, ways). The result of this approach is a collection of simulations that can suggest a range of possible forecasts. Such ensemble forecasts allow the calculation of probabilities: if half

---

4   A sounding is a series of observations at increasing height above the surface.

Sixty-hour forecasts of six-hour precipitation

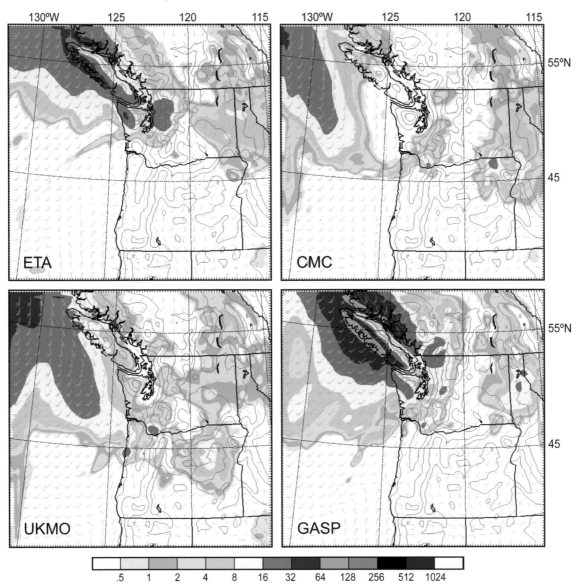

11.20. Sixty-hour forecasts of precipitation based on four different weather-prediction models. The precipitation predictions are in hundredths of an inch and are for a six-hour period ending 4:00 AM, December 14, 2007, using output from the National Weather Service (ETA), Canadian Meteorological Center (CMC), United Kingdom (UKMO), and the Australian (GASP) models. Note the large differences in precipitation among the various simulations, indicating substantial uncertainty in the predictions. These ensemble predictions are available on the University of Washington ensemble prediction Web site (http://www.atmos .washington.edu/~ens/uwme.cgi).

of the forecasts predict rain at Spokane and the rest do not, then a probability of 50 percent might be appropriate. Such ensemble systems even allow meteorologists to predict forecast accuracy. If all the forecasts in the ensemble are providing essentially the same prediction, then one can have more confidence in the predictions. If the forecasts vary substantially, prediction uncertainty is large.

A high-resolution ensemble forecasting system has been running at the University of Washington since 2000, and the National Weather Service has developed a national version as well. An example of how the various ensemble members can differ is illustrated in figure 11.20. The variations among these forecasts are substantial, indicating considerable uncertainty. As computer power increases and the ability to run many forecasts is enhanced, ensemble prediction and probabilistic weather prediction will become the key component of the weather forecaster's toolkit. But even when this technology is perfected, one major challenge remains: learning how to communicate this new kind of weather forecast and educating the public about its benefits.

12

THE EVOLVING WEATHER
OF THE PACIFIC NORTHWEST

||||||||||||||||||||||||||||||||||||||||||||||||||||||||||||||||||||||||||||||||||||||||||||||||||||||||||||||||||||||||||||||||||||||||||||||||||||||||||||||||||||||||||||||||||||||||||||||||

CLIMATE CHANGE IN THE PACIFIC NORTHWEST HAS REACHED THE FRONT PAGE

of local newspapers. The media have given great play to the potential for dramatic

warming due to increasing levels of carbon dioxide and other greenhouse gases as well

as the expectation of major reductions in the mountain snowpacks so crucial for agriculture,

fisheries, and human consumption. A few groups and individuals, including some local

political figures, claim that large declines in snowpack and other serious impacts of warming

have *already* occurred. Others suggest the threat has been exaggerated and that natural

cycles are the cause. Why are climate scientists confident that global warming will have a major impact both globally and locally in the years to come?

RECENT CLIMATE TRENDS

Climate is defined as the average or typical weather for a location or region. Although the definition of climate varies by application, the National Weather Service typically defines climatological conditions using an average taken over the past thirty years. Because the atmosphere has natural year-to-year variability, one must look at changes over several decades to determine if a significant climate trend has occurred. Furthermore, one must be very careful in interpreting climate trends, since they can vary depending on the period considered. In addition, there are a number of causes of climate trends, and unraveling their individual effects can be very difficult. There has been some discussion

and controversy about recent trends in Northwest climate, with the author of this book an active participant.

The University of Washington Climate Impacts Group has analyzed the regional trends in annual temperatures from 1920 to 2000, as shown in figure 12.1. One is struck by the dominance of warming over nearly the entire region, both rural and urban, with many locations heating by 1–3 °F over that period. This warming trend is not the result of a uniform increase in temperature over time. To illustrate this, consider the temperatures averaged over Washington State from 1920 through 2004 (figure 12.2). Although an overall warming trend is evident, there is a large amount of year-to-year variability. While the warmest period was from 1990 to 1999, individual years in the 1930s and the late 1950s were the warmest of the record. Furthermore, little overall warming is evident

12.1. Twentieth-century trends in average annual temperature (1920-2000). Increases (decreases) are indicated with red (blue) dots. The size of the dot corresponds to the magnitude of the change. Data from the U.S. Historical Climate Network. Graphic courtesy of Phil Mote, Climate Impacts Group, University of Washington.

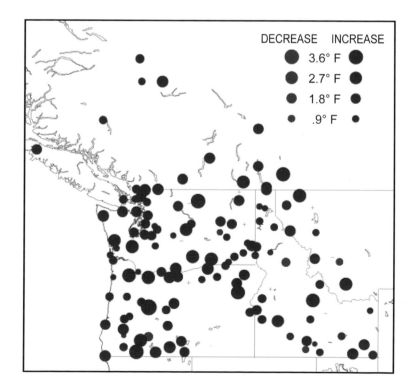

since the mid-1980s, despite the rapid buildup of greenhouse gases during this period. In nearly all respects, the Oregon temperature record is very similar.

Northwest precipitation changes over the same period are presented in figure 12.3, with the precipitation trend expressed in terms of percent-age change over a century. Over the past eighty years, nearly every observing location experienced increasing precipitation, some by 25 percent or more, with the greatest increases over eastern Washington and surrounding areas. Examining the precipitation over all of Washington finds very large year-to-year variability that dwarfs a small

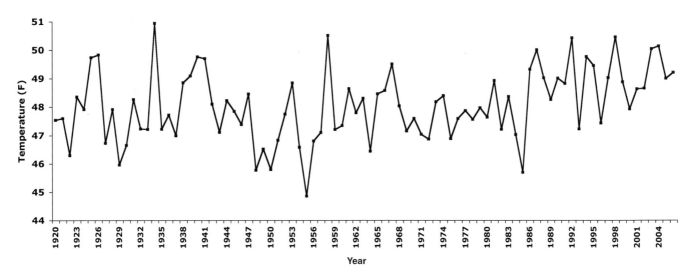

Average Annual Washington State Temperatures

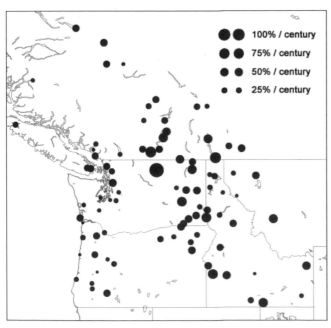

12.2. Temperatures averaged over Washington State from 1920 through 2004. Data from the National Weather Service Climate Division data set.

12.3. Trends in average annual precipitation (1920-2000) expressed as percentage change over a century. Increases (decreases) are indicated with blue (red) dots. The size of the dot corresponds to the magnitude of change. Graphic courtesy of Phil Mote, Climate Impacts Group, University of Washington.

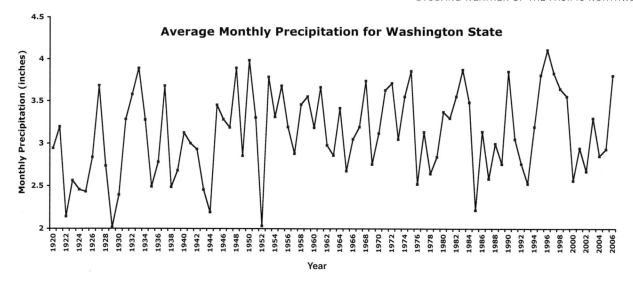

12.4. Average monthly precipitation over Washington State from 1920 through 2006. Although there is a small increasing trend, year-to-year variability is large. Data from National Weather Service Climate Division data set.

increasing trend over the period (figure 12.4). Since the mid-1940s, there has been little precipitation trend over the state.

Although annual precipitation over the Northwest has shown little trend over the past 60–70 years, there has been reduced snowpack in the mountains and less lowland snow over the Puget Sound and Willamette River basins since the mid-1970s compared to the middle of the twentieth century. The health of the mountain snowpack is of vital concern to Northwest residents, since much of our drinking and irrigation water during the summer and early fall is derived from melted mountain snow. Local politicians have cited 50 percent declines in April 1 snowpack in the Cascade Mountains since the 1950s as indications that global warming is real and that society must deal with its causes and impacts. But the reality of the snowpack reduction and its relationship to global warming is complex and often oversimplified in the popular press.

Concerns about snowpack have also been heightened by a dramatic decrease in the size and extent of the high-mountain glaciers of the Northwest. For example, Washington's South Cascade Glacier in the North Cascades has retreated considerably over the past seventy to eighty years, losing approximately half its water content (figure 12.5). In Glacier National Park, located in western Montana, the number of glaciers in the park has dropped from an estimated 150 in 1850 to approximately 35 by 2007. At the current rate of loss, there will probably be no glaciers in the park by the mid-twenty-first century. However, since most of these glaciers have been in retreat for the past 100–150 years, the origin of their loss cannot be solely human-induced global warming. Rather, it appears that the reduction in glacial ice is partly due to the end of the Little Ice Age, a period from approximately 1550 to 1850 AD when much of the northern hemisphere was colder and snowier than before or after. The origin of the Little

(a)

(b)

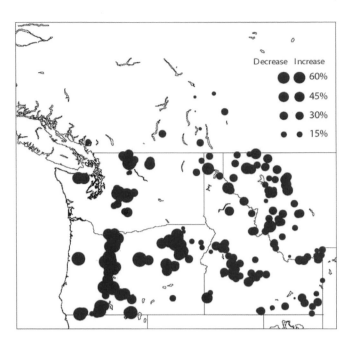

12.6. Trends in April 1 snow water equivalent for the period 1950-2000. Graphic courtesy of Phil Mote, Climate Impacts Group, University of Washington.

12.5. The South Cascade Glacier in the North Cascades retreated dramatically between (a) 1928 and (b) 2000. Photos courtesy of the U.S. Geological Survey.

Ice Age is unclear, with some suggesting a lessening of the output of the sun.

The amount of water in the Cascade Mountain snowpack on April 1 is a key predictor of water availability during the following summer and early fall for east-side agriculture, power generation, fisheries, and other needs. The fifty-year (1950–2000) trend of the snowpack in terms of *snow water equivalent*, the amount of water released by melting all of the accumulated snow and ice, is shown in figure 12.6. Nearly every measurement site showed a decrease in snow water content during that period, with some locations in the Cascades

and coastal mountains experiencing 30–45 percent reductions. But as described below, there is more to this story.

As has been true for temperature, the past changes in snowpack have not been steady or uniform in time. A plot of the April 1 Cascade snowpack since 1940 (figure 12.7) indicates that mountain snowpacks were generally greater from the late 1940s through the early 1970s. After a precipitous decline in the mid-1970s, there has been no apparent trend in the Cascade snowpack during the subsequent thirty years, a period when global warming by human-produced greenhouse gases has been largest. This fact is made clear in figure 12.8, which shows the Cascade Mountain snowpack since 1976 using observations from three elevation ranges (all heights, lowest 25 percent, highest 25 percent). None suggest much of a trend, either increasing or decreasing. It appears that the snowpack trend depends on the period over which it is taken, with recent, short (10- to 30-year) spans showing little trend, while 40- to 60-year trends suggesting significant declines. What can we make of these differences and how can we explain them? What, if any, is the impact of human-forced global warming?

A number of theories have been proposed to explain the trends of temperature and snowpack over the past fifty years or so. One suggestion, of course, is that global warming during this period has caused a significant reduction in the mountain snowpack, and that large year-to-year and decade-to-decade variability makes it difficult to clearly delineate a global-warming signal. Other climate scientists have noted a multi-decade

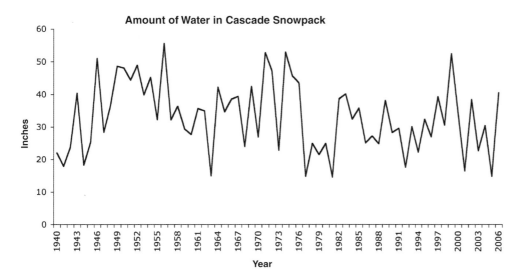

12.7. Water content of Cascade snowpack in Oregon and Washington on April 1. Based on twenty snow courses at a mean elevation of 5,187 feet. Data courtesy of Mark Albright.

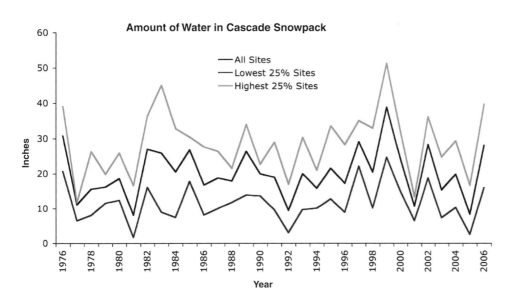

12.8. Average amount of water in the Cascade snowpack for both Oregon and Washington based on snow-course measurements. All sites (86 locations, mean elevation of 4,559 ft) shown in blue, the lowest-elevation 25 percent of sites (21 sites, mean elevation of 2,955 ft) shown in red, and the highest 25 percent of sites (22 locations, mean elevation of 6,098 ft) shown in green. Note that there is little trend in the amount of water in the snowpack at any elevation during the thirty years from 1976 through 2006.

variation in conditions over and within the Pacific Ocean called the *Pacific Decadal Oscillation* (PDO). The PDO has many of the characteristics of the El Niño Southern Oscillation discussed in chapter 11, including varying sea-surface temperatures over the tropical Pacific, but the PDO changes far more slowly. The PDO index, a measure of the strength of the PDO oscillation, appears to vary over a period of roughly thirty years (figure 12.9), with the negative phase associated with cooler and snowier conditions over the Northwest, while the positive phase is typically warmer than normal with less snowpack. The origin of the PDO is not clear, but it is hypothesized that it results from an interaction between the ocean and the atmosphere. Others suggest that the length of the observational record is too short to know whether the PDO is a long-term feature.

As shown in figure 12.9, the PDO was negative from the late 1940s through the mid-1970s, coincident with the period of cooler temperatures and

greater Northwest snowpack. Subsequently, the PDO switched phase and the Northwest transitioned to a warmer, less snowy regime. Might the transition toward increasing temperatures and less snowfall during the last quarter of the twentieth century be at least partly the result of this PDO variability, and not because of global warming due to rising amounts of greenhouse gases? On the other hand, since confidence in the period of oscillation or the nature of the PDO is not great, could greenhouse warming have influenced the PDO index? We simply do not know. The bottom line is that there is no clear indication that global warming induced by mankind has caused a reduction in Northwest snowpack. Ironically, scientists have great confidence that global warming forced by human impacts *will* have a major influence on Northwest weather and climate, including snowpack, during the twenty-first century.

THE FUTURE REGIONAL IMPACT OF HUMAN-INDUCED GLOBAL WARMING

Since the mid-1990s there have been increasing concerns about the regional impacts of global warming, particularly regarding reductions in

12.9. The variation of the Pacific Decadal Oscillation index from 1900 to early 2007. Negative values are associated with cooler and snowier periods over the Pacific Northwest. Graphic courtesy of Nate Mantua, Climate Impacts Group, University of Washington.

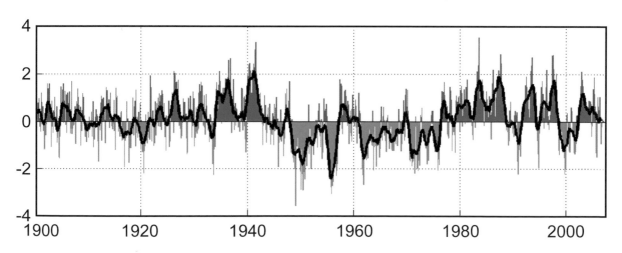

snowpack and water supply. The basic science underlying the prediction of global warming is grounded in strong evidence from both global observations and sophisticated numerical models that rapidly increasing amounts of greenhouse gases, such as carbon dioxide, methane, and nitrous oxide, will warm the earth's surface and the lower portion of the atmosphere (figure 12.10). Mankind's activities, such as the burning of fossil fuels, are the dominant causes of the rapid rise of these gases, which slow the loss of infrared radiation to space—our planet's only way to cool. Thus, greenhouse gases warm the earth in a similar way that blankets warm us at night—by reducing the loss of heat. The most authoritative analysis of the global-warming problem, the product of hundreds of scientists from around the world, has been produced by the Intergovernmental Panel on Climate Change (IPCC) and provides some of the material discussed in this chapter.[1]

General circulation models (GCMs) are major tools used to predict the impact of increasing greenhouse gases. GCMs are highly complex numerical models that simulate both the atmosphere and oceans. These global climate models are very similar to the numerical weather forecasting models that predict weather on a daily basis, except that they allow atmospheric composition to change in time. Modern GCMs can duplicate observed temperature changes over the past century, affording confidence that they can provide realistic predictions of future climate (figure 12.11). Because GCMs are run for decades or longer,

computer resources only allow a relatively low-resolution representation of the atmosphere, so that they cannot resolve the regional terrain of the Northwest, such as the Olympics, the Cascades, or

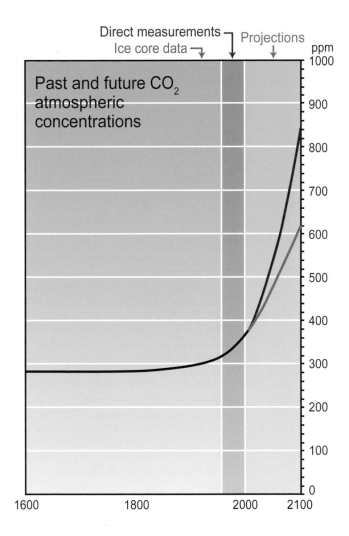

12.10. The amount of carbon dioxide (CO_2), one of the major greenhouse gases, has risen rapidly during the last fifty years and will increase even more quickly during this century. This figure shows two scenarios of carbon dioxide increase based on different assumptions of fossil fuel use: the red line (A2 scenario) represents business as usual, and the green line (B2 scenario) is associated with a significant attempt to restrain greenhouse-gas emission. Graphic from IPCC (2007) report.

---

1   The IPCC Web site is http://www.ipcc.ch.

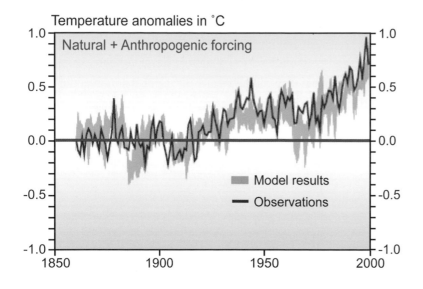

Temperature anomalies in °C

Natural + Anthropogenic forcing

Model results
Observations

12.11. Comparison of observed global temperatures (°C) and the results of a general circulation model. Graphic from IPCC (2007).

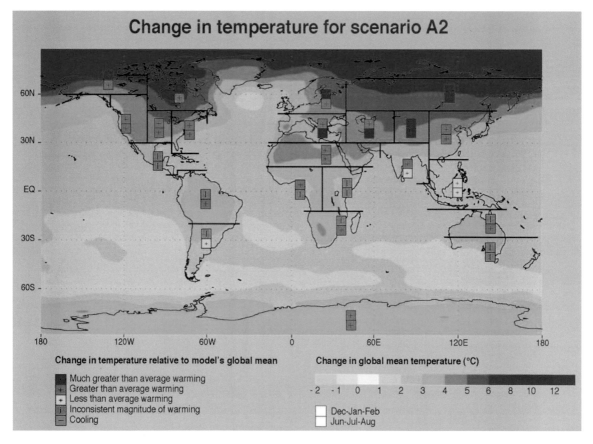

12.12. Temperature change between 2000 and 2100 produced by an average of several general circulation (climate) models. The A2 scenario, which assumes a relatively aggressive increase in greenhouse gases, was used in these calculations. Temperature change is in °C (roughly double that to convert to °F). Note that the greatest warming is in the Arctic and over the interior of the continents. Graphic from IPCC (2007).

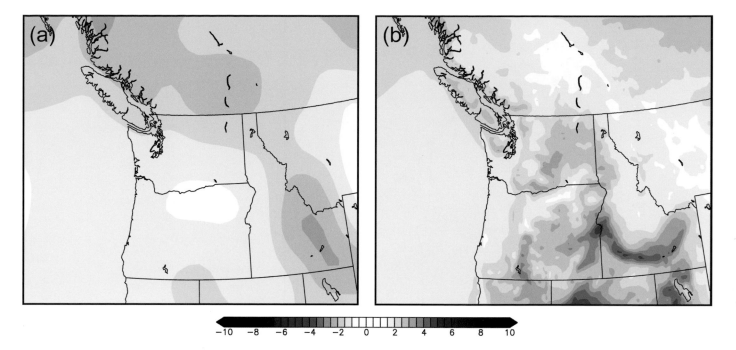

the basin of eastern Washington. Thus, such models only provide a large-scale picture of the implications of global warming, not the local details (figure 12.12). After analyzing a collection of these climate models, the 2007 IPCC report suggests that the planet will warm roughly 2.5–10.5 °F (1.4–5.8 °C) between 2000 and 2100. Furthermore, the warming will not be uniform, with the Arctic heating up the most and land areas warming more than the oceans.

Since regional mountain barriers and land-water contrasts produce a wide variety of local weather features that dominate Northwest weather, one must consider such local effects when predicting the implications of global warming for the region. A major project at the University of Washington has taken on this challenge, running high-resolution regional prediction models over nearly a century into the future. Though these regional models cannot cover the entire planet, they do make use of information from the GCMs

12.13. (a) The change in surface air temperature (°F) from the ECHAM general circulation model between 1995 and 2050 for the winter months (December–February). Note the broad scale of the changes with little local detail in the global model. (b) A high-resolution climate simulation capable of resolving the influence of local terrain. The effects of the mountains, including localized warming on the lower slopes of mountain barriers, are obvious. Graphics courtesy of Eric Salathe and Rick Steed.

noted above. To illustrate the results of such an approach, figures 12.13a and 12.13b show the output from a general circulation model (the ECHAM climate model) and high-resolution simulations using a model applied for local weather prediction (known as the MM5). Both show the changes in surface air temperature (for roughly 6 feet above the surface) between 1995 and 2050 for the winter months of December through February. The global climate model indicates little local detail, with the greatest warming (approximately 3 °F) over Idaho

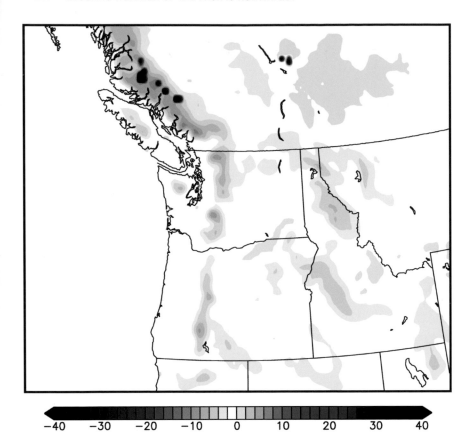

−40   −30   −20   −10   0   10   20   30   40

12.14. Percentage change in the water content of the winter (December–February) snowpack between the 1990s and 2050s. There are large losses on the western slopes of the Cascades and the British Columbia Coast Range. Graphic courtesy of Eric Salathe and Rick Steed.

and southern British Columbia. In contrast, the high-resolution MM5 simulation shows the profound effects of the regional terrain, with 1–2 °F greater warming over the Olympics, Cascades, and eastern Washington than the global model predicts. An interesting detail is that the largest warming is on the lower and middle mountain slopes.

The origin of this enhanced slope warming is probably due to melting of the snow and ice by warming temperatures. Figure 12.14 shows the change in the amount of water in the snowpack during the winter months by 2050 compared to the 1990s. Some of the greatest declines (10–25 percent) are on the lower slopes of the Cascades and their extension into British Columbia, just where local

warming is the greatest. This loss of snow would lead to localized warming because snow cools the lower atmosphere in a number of ways. First, the snow surface is at or below freezing and thus can directly cool the atmosphere above. Second, snow is a good reflector of the sun's rays and an excellent emitter of infrared radiation; both of these effects lead to cooling (figure 12.15). It is not an accident that records for coldest temperatures during the winter normally occur with snow on the ground. Thus, for both reasons the loss of snow on the mountain slopes would produce areas of warming.

Figure 12.16 shows the current and predicted amounts of water in the snowpack at Stampede Pass (4,000 ft) in the central Washington Cascades.

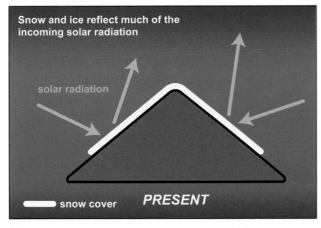

(a)

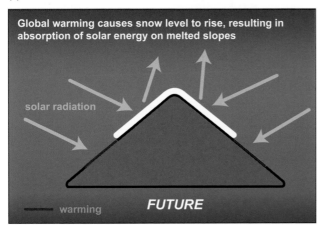

(b)

Between the 1990s and the 2050s, the computer model predicts about a 30 percent reduction in snowpack during the winter and early spring. By the 2090s the loss of snowpack will be extraordinary, with declines of 75 percent or more during the late winter and early spring. The implications of these snowpack reductions are profound: the probable ending of skiing on the lower-elevation slopes such as Snoqualmie Pass and greatly reduced water availability during midsummer and early fall.

12.15. Origin of the enhanced warming on the slopes of Northwest mountain barriers under global warming. (a) Current snow cover on the middle slopes of the mountains during winter and spring months cools the atmosphere, because snow is cold (snow surfaces cannot be warmer than freezing) and because snow reflects solar and emits infrared radiation back to space. (b) Global warming contributes to a higher snow level. Replacing snow with ground or vegetation results in more solar radiation being absorbed and thus warmer temperatures. Also, the cooling effects of the snow surface are lost. Illustrations by Beth Tully/Tully Graphics.

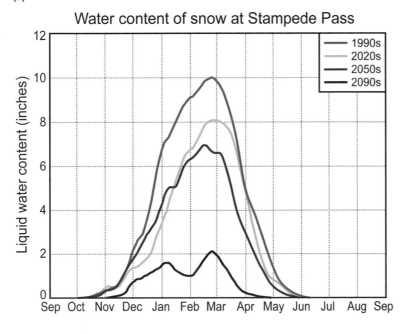

12.16. Snow water content of the snowpack at Stampede Pass, in the central Washington Cascades, for recent and future decades under global warming. These results were produced by the high-resolution MM5 model forced by the ECHAM climate model. Note the extraordinary loss of snowpack between the 2050s and 2090s. Data courtesy of Eric Salathe and Rick Steed. Illustration by Beth Tully/Tully Graphics.

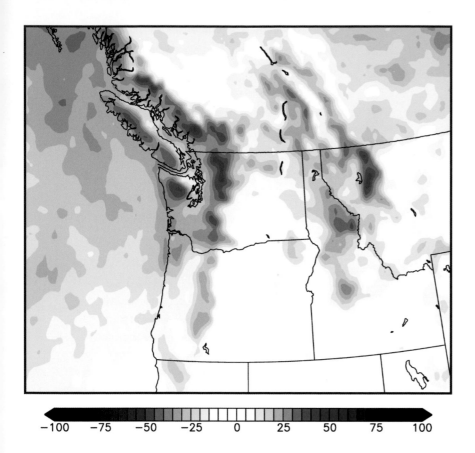

−100   −75   −50   −25   0   25   50   75   100

12.17. Percentage change in the amount of springtime low-level cloud water between the 1990s and 2050s. Such low clouds are mainly stratus and stratocumulus. Note the substantial increases over the western slopes of regional mountain barriers and over portions of western Washington and Oregon. Graphic courtesy of Eric Salathe and Rick Steed.

Regional effects can also *lessen* the impacts of global warming. The high-resolution simulations described above indicate that springtime (March–May) low clouds may actually increase west of the Cascades under global warming (figure 12.17). Since clouds reflect the sun's radiation back to space, this increased cloudiness will lessen daytime warming during the spring over the western lowlands. In addition, global warming will increase low marine clouds over the Pacific as well as the influx of cool, marine air during the spring. Such a seasonal inflow of cool, cloud-laden air is similar to that observed over coastal central and northern California, where late spring and summer are often cloudy and cool. The enhanced spring influx of marine air is caused by changes in the regional pressure distribution: heating of the continental interior by global warming produces a pressure fall in the interior relative to the high-pressure area over the Pacific.[2] A larger onshore pressure difference (with higher pressure offshore) then results in more cool, cloudy Pacific air invading the western side of the Northwest.

Regional climate simulations suggest that local changes in precipitation appear to be very small under global warming, with slight increases in annual precipitation and the greatest enhance-

---

2   Warming causes the air to become less dense. Less air above a point causes surface pressure to fall, since surface pressure is the weight of the air above.

ments during the late fall. Although annual pre-cipitation totals do not appear to change greatly under global warming, the type of precipitation evolves substantially during the winter, with snow replaced by rain at lower and intermediate elevations.

Although the above estimates of the regional effects of global warming are probably the best available, considerable uncertainty exists. Both climate and regional weather-prediction models have weaknesses that will be slowly addressed in the future. Uncertainty also exists regarding the future levels of greenhouse gases and whether people will reduce emissions using new tech-nologies or changed lifestyles. Even if the level of greenhouse gases remained the same as today, a highly improbable scenario, considerable warming would still occur as the atmosphere "catches up" to the warming effects of the greenhouse-gas levels already present in the atmosphere.

13

READING THE PACIFIC NORTHWEST SKIES

||||||||||||||||||||||||||||||||||||||||||||||||||||||||||||||||||||||||||||||||||||||||||||||||||||||||||||||||||||||||||||||||||||||||||||||||||||||||||||||||||||||

BEFORE THE DAYS OF MODERN METEOROLOGICAL TECHNOLOGY, THE CHANGING SKY

offered many clues regarding current and future weather. This is doubly true today, since

cloud interpretation is greatly informed by modern knowledge of the structure and evolution

of weather systems. This chapter gives an illustrated tutorial on Pacific Northwest skies and

what they can tell a perceptive observer.

A specific progression of clouds often signals the approach of fronts and low-pressure

storm systems. The first sign is usually the arrival of wispy *cirrus* clouds made up of millions

of tiny crystals (figure 13.1a). These ice clouds, typically located 20,000 feet or more above sea

level, are caused by the rising motion preceding major weather systems. At first, one sees only

(a)

(b)

13.1. (a) The initial rising motion from approaching weather systems can produce cirrus clouds, which are composed of ice crystals. (b) Ice crystals falling from cirrus clouds into a region of differing winds can produce curved features called mares' tails. Photos courtesy of Art Rangno.

scattered cirrus, sometimes with falling ice crystals that form curved features called *mares' tails* (figure 13.1b). As the weather system moves closer, the cirrus spreads and thickens into a layer covering the sky, known as *cirrostratus* (figure 13.2). Such clouds often produce a halo or ring around the sun or moon due to the bending (or refraction) of light in their six-sided ice crystals. Cirrus and cirrostratus are no reason to turn around during a day hike or similar recreation, since inclement weather is usually six to eighteen hours away, and perhaps the

13.2. A thin layer of cirrus clouds, known as cirrostratus, is often associated with a ring or halo around the sun or moon. Photo courtesy of Art Rangno.

weather system might pass to the north or south. But for a longer trip, preparation for a potential turn to the worst is warranted.

Another early sign of a turn in the weather can be found watching the crests of local mountains, and in particular the tops of major volcanic peaks. Increasing wind and relative humidity associated with approaching weather systems can produce *cap clouds* over the mountain crests (figure 13.3).

A related sign is the development of the lens-shaped lenticular clouds described in chapter 8 (figure 13.4). Such lenticular clouds indicate that winds aloft are strengthening to 20 miles per hour or more and that the atmosphere is getting close to saturation, both indicative of an approaching weather system. So if Mount Rainier, Mount Hood, or one of the other high peaks develops a cap or lenticular cloud, it may be time to check the forecast or the latest weather reports.

If a significant weather system continues to draw near, the cirrus clouds will thicken and lower into *altostratus*, ice and liquid water clouds that

13.3. A cap cloud over Mount Rainier or some other major peak can indicate the approach of a Pacific weather system six to eighteen hours in the future. Photo courtesy of Art Rangno.

are found 6,000–20,000 feet above sea level (figure 13.5). Altostratus clouds leave only a dimmed remnant of the moon or sun, and the terms *watery sun* or *watery moon* are often given to the attenuated solar or lunar orbs. Altostratus is a sign that the weather threat is serious and that rain or snow is perhaps three to six hours away. When altostratus takes over the sky, it is time to head home.

As a weather system draws near, the clouds will descend and thicken further, obscuring the sun and moon completely. At this point, precipitation is imminent and is often heralded by *nimbostratus*, a thick cloud from which precipitation falls (figure 13.6). Initially, the precipitation may evaporate before it reaches the surface, but one can often tell it is coming by the sudden drop in temperature associated with the evaporating raindrops or snow crystals above. Another sign is the visual character of the base of the cloud: a nonprecipitating cloud has a rather sharp, distinct cloud base (figure 13.6a foreground), while a precipitating cloud base has a blurry, washed-out character (figure 13.6a

13.4. Lenticular clouds over eastern Washington. Such clouds often indicate the approach of a weather system. Photo courtesy of Art Rangno.

background). So if the cloud base gets fuzzy and the temperature drops suddenly, it is time to take cover!

Foreseeing the approach of a storm is useful, but so is knowing the signs that the worst is over. The heaviest precipitation and thickest clouds are usually associated with the frontal cloud band,

13.5. The lowering, thickening layer of altostratus clouds allows only a dim image of the sun or moon to be visible. Bad weather is only hours away when the *watery* sun or moon associated with this cloud are seen. Photo courtesy of Art Rangno.

which is typically 50–150 miles wide. As the frontal band moves east of the region, the sky may stay cloudy for a while, but its character will change. Precipitation lessens or stops, and the sky brightens a bit. Often there is a transition from nimbostratus to stratocumulus, as seen in figure 13.7.

Occasional sun breaks appear and the sky starts to open up, with areas of blue between the clouds (figure 13.8). Over the eastern or central United States, the threat of wet weather would be over at this point, but as we shall see, not in the Pacific Northwest.

After cool-season fronts pass east of a Northwest location, there is often an hour or two of brightening skies, but subsequently the weather declines into the region's infamous showers and sun breaks. As noted in chapter 3, cool air usually moves in overhead behind Pacific fronts, the only

(a)

(b)

13.6. (a) Nimbostratus
clouds over Seattle.
Some nonprecipitating
stratocumulus
clouds are apparent
in the foreground.
(b) Altostratus and
altocumulus clouds
overhead with
nimbostratus clouds
in the distance over
Puget Sound. Photos
courtesy of Art Rangno.

(opposite, upper)
13.7. Post-frontal stratocumulus clouds over Seattle.
Photo courtesy of Art Rangno.

(opposite, lower)
13.8. Two hours after a frontal band moved through Puget
Sound, the precipitation temporarily stopped and blue skies
appeared between scattered low clouds.

exception being warm fronts. With relatively mild ocean temperatures, the falling temperatures aloft cause the air to become less stable, often leading to the development of towering cumulus cells, including cumulonimbus, which bring heavy rain and occasionally lightning and thunder (figure 13.9). Sometimes the showers can be heavy, and strong winds often precede or accompany the precipitation. Away from the mountains, these convective showers are relatively brief, lasting ten to fifteen minutes at most. So if you are out gardening and a postfrontal shower threatens, take a coffee break and by the time you are done the showers will probably be over. The approach of such post-frontal showers is often heralded by increasing high clouds and a surge of wind produced by the precipitation-driven downdrafts in the cumulus towers. The sudden onset of such shower-induced winds can be a major issue over the water, particularly for small sailboats. In the mountains or their foothills, these showers can continue for extended periods, as air moving up the terrain continuously forces new convective clouds and showers to form.

The pictures and guidance given above are typical for weather systems during the cool season from late October through early spring. During the later spring through early fall, the cloud systems associated with storms often have a smaller

horizontal scale, and the cloud progression can occur more rapidly. For example, on October 16, 2007, a weak weather system approached western Washington. The progression of cirrus to nimbostratus occurred in only a few hours, with the entire cloud evolution apparent in the western skies that morning (figure 13.10).

Sometimes, signs of changing weather are apparent closer to the ground. For example, it is not without some irony that very dense fog in the morning is often a sign of favorable weather ahead. Such fog limits visibility to tens or hundreds of feet (figure 13.11), making driving dangerous, muffling sound, and producing a dreamlike and somewhat disorienting state. Dense fog is a good omen because it is often associated with high pressure and a lack of clouds aloft. To produce such dense fog at low levels requires intense surface cooling that is usually associated with large emission of infrared radiation to space, which is only possible if the atmosphere above is nearly cloud-free and relatively dry. Furthermore, winds must be light; otherwise, warm air aloft would be mixed down to the surface by the turbulence accompanying stronger winds. In short, if you see very dense fog in the morning, expect an early clearing of the skies and fine weather by lunchtime.

During the summer, the most significant weather feature is the onshore push of marine air that brings substantial cooling and often low clouds (discussed in chapter 7). Watching the sky can sometimes provide signs of an incipient event. Marine pushes are often initiated by the approach of an upper-level disturbance, which is usually signaled by increasing clouds aloft. For example, the development of altocumulus castellanus—middle-level cumulus clouds that look like turrets

(a)

(b)

(opposite)

13.9. After a Pacific front passes through the region, the atmosphere often becomes unstable, leading to the Northwest favorite of showers and sun breaks. (a) A cumulonimbus cloud near Salem, Oregon. Photo courtesy of Art Rangno. (b) A large cumulonimbus over the Cascade foothills with a well-pronounced anvil. The western slopes of the Cascades often focus and maintain the postfrontal instability showers.

(a)

(b)

13.10. The progression from cirrus to nimbostratus to the southwest of Seattle was apparent on October 16, 2007. Only a few hours' warning was given by the evolving cloudscape.

13.11. Dense fog at low levels is often a sign of clear skies aloft and the potential for sun later in the day. (a) Morning dense fog developed over Seattle's Capitol Hill on December 28, 2004. Photo courtesy of Tom Harpel. (b) Thick fair-weather fog is often observed during the early to mid-autumn over the Northwest. This image was taken over northeast Seattle on October 13, 2007. The sun broke through a few hours later with temperatures rising into the mid-60s °F.

or castles in the sky—is a potent warning of a marine push six to eighteen hours in advance (refer back to figure 7.5).

A substantial decline in visibility often accompanies the beginning of a marine push. As described in chapter 7, most marine pushes are preceded by warmer than normal conditions that are produced by offshore and downslope winds. Such offshore flow is very dry and causes water-attracting atmospheric particles to shrink, producing fine visibility. So if you are positioned somewhere west of the Cascade crest during the late spring and summer and the mountains to your east are sharp, well defined, and seemingly only a few

miles away, you can bet that offshore flow is occurring and the weather is entering a warming and improving trend. On the other hand, if you are enjoying sunny, warm weather and suddenly the visibility starts to decline (figure 13.12), the initial stage of a marine push is probably underway. When winds switch from offshore to onshore, the relative humidity of the air rapidly increases, allowing water-attracting particles to grow, thus reducing visibility. Furthermore, breaking waves along the coast eject saltwater droplets into the air, which subsequently dry to produce many tiny salt particles. These particles grow large in humid marine air, contributing to deterioration in visibility.

13.12. During the day of a summer onshore push, the exceptional visibility of the previous day declined, so that the Cascade Mountains behind Everett, Washington, were just barely visible.

13.13. Looking east from north Seattle, the Cascades, roughly 40 miles to the east, appear to be only a short distance away. This situation occurred on October 16, 2007, as low pressure associated with a Pacific weather system was approaching, causing moderate offshore flow. Also note the blue skies associated with air descending the western slopes of the mountains.

13.14. (a) Whitecaps can develop over water when winds reach 15–20 miles per hour. This picture was taken off East Sooke Harbor park near Victoria, British Columbia. Photo courtesy of Chris Campbell. (b) An eastward view of Seattle's Evergreen Point (520) bridge on January 7, 2007. Waves on the southern side indicate southerly winds, and wind-driven whitecaps suggest winds over 30 miles per hour. Photo courtesy of Greg Barnes.

(a)

(b)

Ironically, during the fall and winter months a large improvement in the clarity of the mountains to the east can be a sign of *bad* weather ahead. Fronts and storms are associated with low pressure, so as inclement weather approaches the Northwest a pressure difference develops between higher pressure over the interior and lowering pressure over the coastal zone. This pressure difference forces offshore flow from the dry interior that warms further as it sinks over the western slopes of the Cascades. As noted above, such offshore flow causes atmospheric particles to shrink, improving visibility. In addition, a zone of blue skies is often seen over the western slopes of the mountains where the air is being warmed and dried by descent (this is visible in figure 13.13). Simultaneous with the downslope clearing, clouds stream in from the west as the storm nears; eventually, the incoming storm clouds invade, resulting in overcast and precipitating conditions.

Watching the local waters also reveals a great deal about the weather. As wind increases, waves form on Lake Washington, Puget Sound, and other local bodies of water. By the time the wind strengthens to roughly 15–20 miles per hour, whitecaps develop; and if the wind reaches 30–35 miles per hour, white foam is blown off the crests of the waves (figure 13.14). Local floating bridges have a profound effect on the waves, calming the waters for hundreds of feet on the downwind side. Thus, these bridges act as large but effective wind vanes, with waves on the side facing the wind. Figure 13.14b shows a rather extreme example that occurred on January 7, 2007, when winds from the

13.15. Wind gusts over water are often visible because of the associated darkening of the water, known as cat's-paws.

south gusted to 30–40 miles per hour over land and 40–50 miles per hour over the water.

One can often see the effects of wind gusts over the water by the telltale darkening of the water's surface (figure 13.15). Known as *cat's-paws*, this zone of darker water is the result of waves on the water's surface produced by stronger winds. This corrugation of the surface causes light impinging on the water to be scattered in all directions, lessening the amount of light reaching your eyes. So if you are on a sailboat and see a dark area approaching, better hold on or watch for a swinging sail boom. Over land, gusts can sometimes be seen approaching in the movement of long grass or in the swaying and leaf movement of trees. Strong gusts also announce themselves in advance by an approaching roar, a feature present in many of the Northwest's great windstorms.

PACIFIC NORTHWEST
WEATHER RESOURCES

Excellent information about Pacific Northwest weather can be found online, in several books, and in a rich body of research literature, all of which are described below. I have also established a Web site for this book, http://www.atmos .washington.edu/~cliff/PNWbook.html, which will include corrections, updates to content and resources, and additional information of interest.

PACIFIC NORTHWEST WEATHER INFORMATION ON THE WEB

Perhaps the richest resource for Pacific Northwest weather information is the explosion of Web sites that provide real-time and climatological data for the region. Other Web sites review major Northwest weather features, such as windstorms or gap winds, or survey the impacts of El Niño or La Niña. Although being a meteorologist helps, with a little practice a layperson can understand most of the charts and images available online.

What Is Happening Right Now?
What Is the Latest Forecast?

For the current weather situation, you will want to look at the latest weather satellite and radar images, scan the surface observations for the area of interest, and peruse the most recent forecasts.

A good place to start is the National Weather Service (NWS) and its Northwest offices (Seattle, Spokane, Portland, Pendleton, Medford, Boise, Pocatello), all of which can be accessed through the NWS Web site (http://www.wrh.noaa.gov). The latest forecast for any location is available, as is an analysis of the current weather situation by the lead forecaster on duty, known as the *forecast discussion*. This analysis is a wonderful way to "get into the head" of the forecaster, allowing you to understand the complexities and uncertainties of the forecast.

The University of Washington's Department of Atmospheric Sciences also provides complete weather information for the region, including high-resolution computer weather predictions (http://www.atmos.washington.edu/weather.html).

Local TV stations all have weather Web sites, but with more limited offerings. I do occasionally visit the Web sites of local TV stations that maintain their own weather radars, since they help fill in the gaps of the NWS network. And, of course, there are Web sites run by large private-sector weather outlets like the Weather Channel (http://www.weather.com).

Using the weather radars, you can determine where it is raining or snowing now and make a short-term extrapolation of where it will be precipitating in the near future. A substantial portion of the region has poor radar coverage (e.g., the coast, the eastern slopes of the Cascades), but if you are lucky to be in an area with coverage, weather radar can be useful in planning daily activities. My favorite radar Web sites follow:

► National Weather Service weather radars (http://radar.weather.gov). High-quality Doppler weather radars located at Camano Island and Spokane, Washington; Portland, Pendleton, and Ashland, Oregon; and Boise and Pocatello, Idaho.

► KGW-TV, Portland (http://www.kgw.com/weather/doppler.html).

► KING5-TV, Seattle (http://www.king5.com/weather/doppler/?seattle). Looks out to the central Washington coast, a region not covered by the NWS weather radar.

► Environment Canada (http://weatheroffice.ec.gc.ca/radar). Four weather radars in British Columbia, with three providing some coverage of the northwestern United States.

Let's consider a specific example. I would like to go on a hike today on the western slopes of the Washington Cascades near Snoqualmie Pass, and I want to get an idea of what weather to expect. The following is the approach I would probably follow:

First, I would examine the animations of the latest visible and infrared satellite imagery for the region to determine the cloud cover and to see what changes are occurring (http://www.wrh.noaa.gov/satellite/).

Next, if there are any clouds, I would view the latest radar images to see if there is any precipitation in the area (http://radar.weather.gov/radar.php?rid=atx).

Then, if things look decent, I would check the latest NWS forecast for the western slopes of the Cascades and then review the forecast discussion, using the University of Washington's Department of Atmospheric Sciences site (http://www.atmos.washington.edu/data/zone_report.KSEW.html and http://www.atmos.washington.edu/data/disc_report.html#ksew) or the Seattle NWS site (http://www.wrh.noaa.gov/sew; select "Zone Forecast" on the left).

I would then probably look at the latest weather observations at some mountain stations, such as Stampede Pass (4,000 ft, 9 miles southeast of Snoqualmie Pass), via the University of Washington's Department of Atmospheric Sciences (http://www.atmos.washington.edu/ data/nw.cgi?what=0; the identifier for Stampede Pass, in the left-hand column, is KSMP). Or I could examine the latest mountain observations at weather observation sites maintained by the Northwest Avalanche Center, which has a sensor at Snoqualmie Pass (http://www.nwac.us/ mtnweather.htm).

Finally, if I planned on being in the mountains for a while, I would check the high-resolution computer forecasts available through the University of Washington's Department of Atmospheric Sciences (http://www.atmos.washington.edu/mm5rt. Interpreting these graphics may be difficult, but not impossible, for interested laypersons.

What Are the Average or Climatological Conditions for a Specific Location?

For a number of reasons, many of us want to know the typical or climatological conditions for a location. What is the average rainfall or maximum temperature for a date? When is the best time of the year to plan that special outdoor party? Where is the driest location west of the Cascades? Massive amounts of such information are available online in convenient formats—which is why such data are not found in long appendices at the end of this book.

Probably the best Web site for Northwest climate information is the Western Region Climate Center (http://www.wrcc.dri.edu/Climsum.html). At this site you can view graphs and tabular data for hundreds of official observation sites around the Northwest, with many records going back decades

or longer. Additional climatological data can be found at Web sites provided by the Office of the Washington State Climatologist (http://www .climate.washington.edu/climate.html) and the Oregon Climate Service (http://www.ocs.orst.edu/ index.html).

Let's consider an example. Suppose I wanted to find the driest time of the year in Salem, Oregon, to hold a wedding. My steps would be the following:

First, I would go to the Western Region Climate Center and select Oregon (http://www.wrcc.dri .edu/summary/Climsmor.html).

Then I would click on the Salem Airport location on the map or from the list on the left (http:// www.wrcc.dri.edu/cgi-bin/cliMAIN.pl?or7500).

Clicking on daily temperature and precipitation on the left-hand side would bring up a graph, which clearly shows that the best period for my wedding is during late July or early August.

What Is the Expected Weather over the Next Few Months?

Although skillful weather prediction is only possible for a week or two in advance, there is some marginal skill in forecasting the general character of the weather during the upcoming months. A few Web sites are dedicated to providing sophisticated climate diagnostics and predictions of short-term climate trends.

Two major sites run by NOAA provide information on recent climate trends and predictions for the future. The Climate Prediction Center (http:// www.cpc.ncep.noaa.gov) furnishes forecasts from six to ten days through approximately one year ahead. Detailed information on El Niño and La

Niña is also available, including predictions for their future evolution. The Climate Analysis Branch of NOAA's Earth System Research Laboratory (ESRL) is a good place to get other climate information (http://www.cdc.noaa.gov).

Additional information on El Niño and La Niña can be found at the excellent Web site provided by the Pacific Marine Environment Lab (http://www .pmel.noaa.gov/tao/elnino/nino-home.html). If your interest is the impacts of climate change on the Pacific Northwest, much information is available from the University of Washington Climate Impacts Group (http://www.cses.washington .edu/cig).

NOAA WEATHER RADIO

Another source of real-time weather information is the National Oceanic and Atmospheric Administration (NOAA) NWS weather radio service. The NWS broadcasts weather forecasts and warnings that can be received on relatively inexpensive receivers (twenty-five to fifty dollars), available at electronics stores. A nice feature of these receivers is their warning capability: if a major storm or life-threatening event is predicted, an alarm on the receiver is triggered.

BOOKS ON PACIFIC NORTHWEST WEATHER

A number of books have been written in total or in part on Pacific Northwest weather. Each has a different emphasis, many are written by non-meteorologists, and some possess significant technical errors. The following list is relatively comprehensive and includes books that are currently out of print, but still available in libraries, on eBay, and from some used book dealers.

Lange, Owen S. 1998. *The Wind Came All Ways*. Vancouver, BC: Environment Canada. Exhaustive description of the winds over the waters of southern British Columbia.

Laskin, David. 1997. *Rains All the Time*. Seattle: Sasquatch Books. Provides colorful descriptions of major historical weather events over the Northwest.

Lilly, Kenneth E., Jr. 1983. *Marine Weather of Western Washington*. Seattle: Starpath. Written by a local NOAA commander, this book, although somewhat out of date, has useful descriptions of local weather features over and near the waterways of western Washington.

Miller, George. 2002. *Pacific Northwest Weather*. Portland: Frank Amato Publications. A wide-ranging review of regional weather, with some stress on Oregon. Written by a retired National Weather Service forecaster.

——. 2004. *Lewis and Clark's Northwest Journey: Weather Disagreeable*. Portland: Frank Amato Publications. Weather stories from the famous 1804–06 expedition.

Pool, Steve, and Scott Sistek. 2005. *Somewhere I Was Right*. Seattle: Classic Day/Peanut Butter Publishing. Pool is a well-known TV weathercaster in Seattle and Sistek has a bachelor's degree in atmospheric sciences from the University of Washington. Reviews several Northwest weather features and includes a description of the fun and travails of being a TV weather personality.

Renner, Jeff. 1992. *Northwest Mountain Weather: Understanding and Forecasting for the Back-country User*. Seattle: Mountaineers Books. This and the next book were written by a well-known Seattle TV weathercaster who received an undergraduate atmospheric sciences degree from the University of Washington.

———. 1993. *Northwest Marine Weather: From Columbia to Cape Scott*. Seattle: Mountaineers Books. Stresses winds over local waterways.

Rue, Walter. 1978. *Weather of the Pacific Coast*. Vancouver, BC: Gordon Soules Book Publishers. Historical and climatological descriptions are its main strengths.

Taylor, George, and Chris Hannan. 1999. *The Climate of Oregon: From Rain Forest to Desert*. Corvallis: Oregon State University Press. Climatological data for Oregon.

Taylor, George, and Raymond Hatton. 1999. *The Oregon Weather Book*. Corvallis: Oregon State University Press. Extensive historical information and descriptions of major Oregon weather events. Taylor was the controversial Oregon State climatologist.

RESEARCH LITERATURE ON PACIFIC NORTHWEST WEATHER

Since the mid-1970s, there has been a considerable amount of research on Pacific Northwest weather by both university and government scientists. These efforts have unraveled many of the mysteries of Northwest weather and are relatively accessible even to those without a formal meteorological education. The research results are found mostly in journals published by the American Meteorological Society (AMS), and they are accessible at major local libraries or online (AMS Journals Online, http://ams.allenpress.com/perlserv/?request=index-html). The most relevant references for each of this book's chapters follow.

1 The Extraordinary Weather of the Pacific Northwest

Church, P. 1962. Seattle: A spectacular and clean air city. *Weatherwise* 15:12–23.

Phillips, E. 1962. Weather highlights of the Pacific Northwest. *Weatherwise* 15:30–38.

2 The Basics of Pacific Northwest Weather

Church, P. E., and T. E. Stephens. 1941. Influence of the Cascade and Rocky Mountains on the temperature during westward spread of polar air. *Bulletin of the American Meteorological Society* 22:25–30.

3 Floods

Colle, B. A., and C. F. Mass. 2000. The 5–9 February 1996 flooding event over the Pacific Northwest: Sensitivity studies and evaluation of the MM5 precipitation forecasts. *Monthly Weather Review* 128:593–617.

4 Snowstorms and Ice Storms

Ferber, G., C. Mass, G. Lackmann, and M. W. Patnoe. 1993. Snowstorms over the Puget Sound lowlands. *Weather and Forecasting* 8:481–504.

5 Windstorms

Lynott, R. E., and O. P. Cramer. 1966. Detailed analysis of the 1962 Columbus Day windstorm in Oregon and Washington. *Monthly Weather Review* 94:105–17.

Mass, C., and G. Ferber. 1990. Surface pressure perturbations produced by an isolated mesoscale topographic barrier, part I: General characteristics and dynamics. *Monthly Weather Review* 118:2579–96.

Read, Wolf. The storm king: Some historical weather events in the Pacific Northwest. http://www.climate.washington.edu/stormking.

Reed, R. 1980. Destructive winds caused by an orographically induced mesoscale cyclone. *Bulletin of the American Meteorological Society* 61:1346–55.

Reed, R. J., and M. Albright. 1985. A case study of explosive cyclogenesis in the eastern Pacific. *Monthly Weather Review* 114:2297–2319.

Steenburgh, W. J., and C. F. Mass. 1996. Interaction of an intense extratropical cyclone with coastal orography. *Monthly Weather Review* 124:1329–52.

6 Sea Breezes, Land Breezes, and Slope Winds

Buettner, K., and N. Thyer. 1962. Valley winds in Mt. Rainier National Park. *Weatherwise* 15:46–53.

Cross, C. 1950. Slope and valley winds in the Columbia River valley. *Bulletin of the American Meteorological Society* 31:79–84.

Doran, J. C., and S. Zhong. 1994. Regional drainage flows in the Pacific Northwest. *Monthly Weather Review* 122:1158–67.

Johnson, A., and J. J. O'Brien. 1973. A study of an Oregon sea breeze event. *Journal of Applied Meteorology* 12:1267–83.

Mass, C. 1982. The topographically forced diurnal circulations of western Washington State and their influence on precipitation. *Monthly Weather Review* 110:170–83.

Staley, D. O. 1957. The low-level sea breeze of northwest Washington. *Journal of Meteorology* 14:458–70.

———. 1959. Some observations of surface-wind oscillations in a heated basin. *Journal of Meteorology* 16:364–70.

7 Coastal Weather Features

Bond, N., C. Mass, and J. Overland. 1996. Coastally trapped southerly transitions along the U.S. West Coast during the warm season, part I: Climatology and temporal evolution. *Monthly Weather Review* 124:430–45.

Chien, F. C., and C. F. Mass. 1997. A numerical study of the interaction between a warm-season frontal system and the coastal mountains of the western U.S., part II: Evolution of a Puget Sound Convergence Zone. *Monthly Weather Review* 125:1730–52.

Chien, F. C., C. F. Mass, and Y.-H. Kuo. 1997. A numerical study of the interaction between a warm-season frontal system and the coastal mountains of the western U.S., part I: Prefrontal pressure ridge, onshore push, and alongshore southerlies. *Monthly Weather Review* 125:1705–29.

Colle, B. A., and C. F. Mass. 2000. High-resolution observations and numerical simulations of easterly gap flow through the Strait of Juan de Fuca on 9–10 December 1995. *Monthly Weather Review* 128:2363–96.

Hermann, A. J., B. Hickey, C. Mass, and M. Albright. 1990. Coastally trapped atmospheric gravity currents in the Pacific Northwest and their oceanic response. *Journal of Geophysical Research* 95:13169–93.

Mass, C., 1980. The Puget Sound Convergence Zone. *Weatherwise* 33:272–74.

———. 1981. Topographically forced convergence in western Washington State. *Monthly Weather Review* 109:1335–47.

———. 1987. The "banana belt" of the southern Oregon coast. *Weather and Forecasting* 2:187–98.

Mass, C., M. Albright. 1987. Coastal southerlies and alongshore surges of the west coast of North America: Evidence of mesoscale topographically trapped response to synoptic forcing. *Monthly Weather Review* 115:1707–38.

Mass, C., M. Albright, and D. J. Brees. 1986. The onshore surge of maritime air into the Pacific Northwest: A coastal region of complex terrain. *Monthly Weather Review* 114:2602–27.

Mass, C., and N. Bond. 1996. Coastally trapped wind reversals along the U.S. West Coast during the warm season, part II: Synoptic evolution. *Monthly Weather Review* 124:446–61.

Mass, C., S. Businger, M. Albright, and Z. Tucker. 1995. A windstorm in the lee of a gap in a coastal mountain barrier. *Monthly Weather Review* 123:315–31.

Mass, C., and D. Dempsey. 1985. A topographically forced convergence line in the lee of the Olympic Mountains. *Monthly Weather Review* 113:659–63.

Mass, C., and G. Ferber. 1990. Surface pressure perturbations produced by an isolated mesoscale

topographic barrier, part I: General characteristics and dynamics. *Monthly Weather Review* 118:2579–96.

Mass, C., and J. Steenburgh. 2000. An observational and numerical study of an orographically trapped wind reversal along the West Coast of the U.S. *Monthly Weather Review* 128:2363–96.

Overland, J. E., and B. A. Walter Jr. 1981. Gap winds in the Strait of Juan de Fuca. *Monthly Weather Review* 109:2221–33.

Reed, T. J. 1931. Gap winds of the Strait of Juan de Fuca. *Monthly Weather Review* 59: 373–76.

8 Mountain-Related Weather Phenomena

Cameron, D. C. 1931. Easterly gales in the Columbia River Gorge during the winter of 1930–1931—some of their causes and effects. *Monthly Weather Review* 59:411–13.

Cameron, D. C., and A. B. Carpenter. 1931. Destructive easterly gales in the Columbia River Gorge. *Monthly Weather Review* 64:264–67.

Colle, B. A., and C. F. Mass. 1998. Windstorms along the western side of the Washington Cascade Mountains, part I: A high resolution observational and modeling study of the 12 February 1995 event. *Monthly Weather Review* 126:28–52.

———. 1998. Windstorms along the western side of the Washington Cascade Mountains, part II: Characteristics of past events and three-dimensional idealized simulations. *Monthly Weather Review* 126:53–71.

Krist, Gary. *The White Cascade: The Great Northern Railway Disaster and America's Deadliest Avalanche.* New York: Henry Holt, 2007.

Mass, C., and M. Albright. 1985. A severe windstorm in the lee of the Cascade mountains of Washington State. *Monthly Weather Review* 113:1261–81.

Mass, C., S. Businger, M. Albright, and Z. Tucker. 1995. A windstorm in the lee of a gap in a coastal mountain barrier. *Monthly Weather Review* 123:315–31.

Reed, R. J. 1981. A case study of a bora-like windstorm in western Washington. *Monthly Weather Review* 109:2383–93.

Sharp, J., and C. F. Mass. 2002. Columbia Gorge flow: Insights from observational analysis and ultra-high resolution model simulation. *Bulletin of the American Meteorological Society* 18:75–79.

——. 2004. The climatological influence and synoptic evolution associated with Columbia Gorge gap flow events. *Weather and Forecasting* 19:970–92.

Steenburgh, J., C. F. Mass, and S. A. Ferguson. 1997. The influence of gaps in a coastal mountain barrier on temperature and snow level. *Weather and Forecasting* 12:208–27.

9 Weather Features of the Inland Northwest

Baker, R., E. Wendell Hewson, N. G. Butler, and E. J. Warchol. 1978. Wind power potential in the Pacific Northwest. *Journal of Applied Meteorology* 17:1814–25.

Cameron, D. 1931. Great dust storm in Washington and Oregon, April 21–24, 1931. *Monthly Weather Review* 64:195–97.

Mass, C., and D. Portman. 1989. The effect of major volcanic eruptions of the last century on surface temperature, pressure and precipitation. *Journal of Climate* 2:566–93.

Mass, C., and A. Robock. 1982. The short-term influence of the Mount St. Helens volcanic eruption

on surface temperature in the northwest United States. *Monthly Weather Review* 110:614–22.

Pitzer, P. 1988. The atmosphere tasted like turnips: The Pacific Northwest dust storm of 1931. *Pacific Northwest Quarterly* 79.2 (April): 50–55.

Robock, A. 1981. The Mount St. Helens volcanic eruption of 18 May 1980: Minimal climatic effect. *Science* 212:1383–84.

Robock, A., and C. Mass. 1982. The Mount St. Helens volcanic eruption of 18 May 1980: Large short-term effects. *Science* 216:595–610.

10 Blue Holes, Flying Ferries, and Tornadoes

Huse, T. 1995. *Washington State Tornadoes*. NWS Technical Memorandum PB96–107024. July. NWS Western Region, Salt Lake City, UT.

11 The Challenge of Pacific Northwest Weather Prediction

Mass, C., M. Albright, D. Ovens, R. Steed, M. MacIver, E. Grimit, T. Eckel et al. 2003. Regional environmental prediction over the Pacific Northwest prototype. *Bulletin of the American Meteorological Society* 84:1353–66.

Westrick, K., C. Mass, and B. Colle. 1999. Is meteorological radar useful for quantitative precipitation estimation over the western U.S.? *Bulletin of the American Meteorological Society* 80:2289–98.

12 The Evolving Weather of the Pacific Northwest

Intergovernmental Panel on Climate Change (IPCC). 2007. *Climate Change 2007: The Physi-*

cal Science Basis. IPCC Fourth Assessment Report. http://www.ipcc.ch/ipccreports/ar4-wg1.htm.

Mantua, N. J., S. R. Hare, Y. Zhang, J. M. Wallace, and R. C. Francis. 1997. A Pacific interdecadal climate oscillation with impacts on salmon pro-duction. *Bulletin of the American Meteorological Society* 78:1069–79.

Mote, P. W., A. F. Hamlet, M. P. Clark, and D. P. Lettenmaier. 2005. Declining mountain snow-pack in western North America. *Bulletin of the American Meteorological Society* 86:39–49.

INDEX